FROGS
AND
TOADS
OF ALABAMA

Philip Henry Gosse as a young man of twenty-nine, the year of his return to England from Alabama, painted by his brother, William Gosse. (1839, watercolor on ivory, courtesy of the National Portrait Gallery—London)

Philip Henry Gosse (1810–1888) was an English naturalist and illustrator who spent eight months of 1838 on the Alabama frontier, teaching planters' children in Dallas County and studying the native flora and fauna. Years later, he published the now-classic *Letters from Alabama, Chiefly Relating to Natural History*, which draws on detailed reports sent back from his visit and features twenty-nine important black-and-white illustrations. During his time in Alabama, Gosse also produced forty-nine watercolor plates of various plant and animal species, mainly insects, now available in *Philip Henry Gosse: Science and Art in "Letters from Alabama" and "Entomologia Alabamensis."*

The Gosse Nature Guides are a series of natural history guidebooks prepared by experts on the plants and animals of Alabama and designed for the outdoor enthusiast and ecology layperson. Because Alabama is one of the nation's most biodiverse states, its residents and visitors require accurate, accessible field guides to interpret the wealth of life that thrives within the state's borders. The Gosse Nature Guides are named to honor Philip Henry Gosse's early appreciation of Alabama's natural wealth and to highlight the valuable legacy of his recorded observations. Look for other volumes in the Gosse Nature Guides series at www.uapress.ua.edu.

FROGS

AND

TOADS

OF ALABAMA

CRAIG GUYER AND MARK A. BAILEY

WITH LINE DRAWINGS BY **CLAIRE C. FLOYD**

THE UNIVERSITY OF ALABAMA PRESS   TUSCALOOSA

The University of Alabama Press
Tuscaloosa, Alabama 35487-0380
uapress.ua.edu

Typeface: Scala Pro

Cover image: Pine Barrens Treefrog; courtesy of Aubrey M. Heupel
Cover design: Michele Myatt Quinn

Cataloging-in-Publication data is available from the Library of Congress.
ISBN: 978-0-8173-6066-5
E-ISBN: 978-0-8173-9429-5

WE DEDICATE THIS VOLUME TO THE MEMORY OF ROBERT H. MOUNT, the author of the original field guide to amphibians and reptiles of Alabama and an important collaborator in our efforts to update that volume. Bob's original book was widely acclaimed for its scientific rigor and its accessibility to novices. He honored us by requesting that we revise the volume and was persistent in his efforts to assist us in finishing that project. Unfortunately, Bob passed away on September 10, 2017. His passing prevented his involvement as a coauthor, as he had been on the reptile volumes. Nevertheless, his influence on us was great and is present throughout this volume. Those who know Bob and his work will recognize his imprint, and we hope our efforts kindle the interest in nature that characterized Bob's life.

In going, I had heard, from a wet marshy place beside the road, a continued and most deafening shrieking, extremely shrill and loud. When I came to the place in returning, the noise was still kept up, and my curiosity was much excited. I watched, and had reason to believe it was produced by a small dusky species of frog, for, on approaching the spots whence it proceeded, it instantly ceased, at least there, and two or three of these frogs would dash into the water and dive. Wishing much to witness the act of uttering the sound, (which was no easy matter, for, as I have said, it ceased on the approach of a foot,) I crept cautiously to the edge of one of the little pools, in which I saw two or three frogs: they were very shy, and kept under water, but I waited patiently, quiet and motionless, a long time, taking care not to stir hand or foot. At length, one of them, taking courage, raised his head and half of his body out of water, sitting up, as it were, and resting on the toes of his forefeet, and thus uttered the piercing shriek, which had a kind of cracked or ringing sound, somewhat like that of a penny trumpet. When about to cry, he first fluttered the skin of the throat a few times, then suddenly inflated it, till it was like a blown bladder, perfectly round, as big as his head, which continued so all the time of the shriek, about four or five seconds. I saw him do it many times close to my feet: it was a very singular sight.

—Philip Henry Gosse, *Letters from Alabama*, Letter V, Dallas County, June 16, 1838

Contents

Ranid Frogs

Great Caribbean Landfrogs

Treefrogs

Abbreviations

Throughout this book, various agencies, programs, and legislation are frequently represented as acronyms with which the reader should become familiar.

ADCNR Alabama Department of Conservation and Natural Resources
ANHP Alabama Natural Heritage Program
AUM Auburn University Museum of Natural History
CITES Convention on International Trade in Endangered Species
ESA Endangered Species Act of 1973, as amended
USFS US Forest Service
USFWS US Fish and Wildlife Service

FROGS
AND
TOADS
OF ALABAMA

Introduction

This book is designed to update information on the anurans (frogs and toads) of Alabama covered in Robert H. Mount's (1975) comprehensive volume describing the amphibians and reptiles of the state. Mount's work was seminal in summarizing the taxonomy and life history of one of the most species-rich regions in the northern hemisphere and has served as the key reference on the anuran fauna of the state since its publication. The volume was unusual in documenting variation within each species across the state, an important feature because of the frequent subspecific variation of Alabama's biota associated with the state's unusual position at the southern end of the Appalachian Mountains, its dissection by several rivers, and its association with the mild climate of the Gulf of Mexico. This varied topography and climate result in changes in color patterns, reproductive seasons, and likely call dialects of frogs and toads. Additionally, Mount's volume was exemplary for museum-oriented works in its foundation in the research collections, accumulated largely by Jack Mecham, George Folkerts, Jim Dobie, Mount, and their students at Auburn University. Therefore, the descriptive accounts of subspecific variation were accompanied by the types of morphological data needed to support taxonomic decisions made at the time.

Much has happened in organismal biology since 1975. Significant field studies, especially of Alabama's threatened and endangered species, have been performed. The field of systematics has re-emerged as a primary goal of biological sciences, and this has been coupled with a healthy debate on species concepts (e.g., Frost and Hillis 1990). This debate has expanded the focus of studies of speciation, changing them from tests of reproductive isolation (e.g., Mecham 1960) to discovery of diagnostic features indicative of unique lineages on phylogenetic trees (e.g., Frost et al. 2006). These changes have increased the known diversity of the state and pointed out new directions for research that are likely to continue to expand Alabama's known anuran fauna.

Because knowledge accumulates, descriptions of important aspects of the life history of each species have increased. For that reason, the size of an updated volume on the entire herpetofauna of Alabama would be quite large. Therefore, we have separated the update of Alabama's herpetofauna into a series of volumes, and this one represents the first focused on amphibians. Our target audience remains the same as that for Mount (1975). We aim to enlighten people who are interested in the natural history of their local biota because we know these people will develop responsible attitudes toward the role that humans play in sustaining the earth's biotic diversity. Moreover, those with knowledge of natural history and a willingness to experience nature have a vast new world full of opportunities for soul-enriching experiences that we have had as biologists and hope to generate for others. This publication is designed to provide a basis for understanding the extant anurans of Alabama. It is a compromise of sorts in that it was prepared for use by both the layperson and the serious student of southeastern herpetology as well.

THE FROG AND TOAD FAUNA OF ALABAMA

Indigenous Species

The following classification scheme presents our organization of the native frogs and toads of Alabama. These taxa are thought to have evolved within the state or to have dispersed there without assistance of humans. Changes to systematic biology since the publication of Mount (1975) have generated a growing number of taxonomic groupings that are monophyletic (groups in which members are all more closely related to each other than any member is to a species outside of the group) and a desire for restricting the proliferation of named groups associated with monophyletic taxonomies. To reach these goals, we have adopted some of the philosophy argued by de Queiroz and Gauthier (1992), who advocate reducing the reliance on taxonomic levels of the Linnaean hierarchy in favor of generating indented lists of increasingly more restricted monophyletic groups. Even in such taxonomies, species are identified with binomials: the genus name identifying a group of closely related species, and the specific epithet identifying a particular one of those species. The species name includes both the genus and the specific epithet, simultaneously generating a unique name for each species and identifying it as part of a more inclusive taxonomic group. In addition to this convention, we retain the level of family as a useful taxonomic category because this level is so heavily entrenched in the taxonomic literature and because the content of anuran families has remained relatively consistent. We have avoided use of terms associated with levels of the Linnaean hierarchy above the level of the family because these vary substantially among schemes, and the choice of a term for these levels (e.g., superfamily versus suborder) is a matter of personal choice rather than providing any increased understanding of biology.

Our classification scheme uses the format of an indented list, starting with the group Anura, the radiation that contains all living and fossil frogs and toads. At the first level of indentation are the seven living families known from Alabama. These are listed in alphabetical order because attempts to discover the evolutionary relationships of these families have not converged on a consistent phylogenetic tree. Within families we list species and subspecies in alphabetical order.

Anura
  Bufonidae
     Anaxyrus americanus americanus—Eastern American Toad
     A. fowleri—Fowler's Toad
     A. quercicus—Oak Toad
     A. terrestris—Southern Toad
  Hylidae
     Acris crepitans—Eastern Cricket Frog
     A. gryllus—Southern Cricket Frog
     Dryophytes andersonii—Pine Barrens Treefrog
     D. avivoca—Bird-voiced Treefrog
     D. chrysoscelis—Cope's Gray Treefrog
     D. cinerea—Green Treefrog
     D. femoralis—Pine Woods Treefrog
     D. gratiosa—Barking Treefrog
     D. squirella—Squirrel Treefrog
     Pseudacris brachyphona—Mountain Chorus Frog
     P. crucifer—Spring Peeper
     P. feriarum—Upland Chorus Frog
     P. nigrita—Southern Chorus Frog
     P. ocularis—Little Grass Frog
     P. ornata—Ornate Chorus Frog
  Microhylidae
     Gastrophryne carolinensis—Eastern Narrow-mouthed Toad
  Ranidae
     Lithobates areolatus circulosus—Northern Crawfish Frog

L. capito—Gopher Frog
L. catesbeianus—American Bullfrog
L. clamitans—Green Frog
L. grylio—Pig Frog
L. heckscheri—River Frog
L. palustris—Pickerel Frog
L. sevosus—Dusky Gopher Frog
L. sphenocephalus sphenocephalus—Florida Leopard Frog
L. sphenocephalus utricularius—Coastal Plains Leopard Frog
L. sylvaticus—Wood Frog
Scaphiopodidae
Scaphiopus holbrookii—Eastern Spadefoot

Introduced Species

Because of increased trade in vertebrates, the establishment and expansion of non-indigenous species has become an increasing problem in maintaining native North American faunas (Romagosa et al. 2009). Much of the trade in North American amphibians passes through South Florida, where individuals that escape captivity or are intentionally released have established populations of a growing number of species. The majority of these have not spread from the areas occupied by the founding populations (Meshaka et al. 2004). However, in a few cases these taxa have expanded rapidly and extensively from the founding population and concern has emerged that these taxa might disrupt native communities. For example, the Cane Toad (*Rhinella marina*) is now established throughout most of Florida and some ecological niche models predict that it will spread across the Gulf Coast of Alabama (Holcombe et al. 2008). This species reaches dense populations, is a voracious predator, and produces noxious defensive toxins. Therefore, Cane Toads appear capable of altering populations of native amphibians by altering their food resources, preying directly upon them, or modifying the abundance of their native predators (Crossland et al. 2008; Greenlees et al. 2006; Phillips et al. 2003).

Mount (1975) listed no non-indigenous anuran species for the state of Alabama. Since that publication, three species—the Greenhouse Frog (*Eleutherodactylus planirostris*), a species native to Cuba; the Rio Grande Chirping Frog (*E. cystignathoides*), a species native to southern Texas and northern Mexico; and the Gulf Coast Toad (*Incilius nebulifer*), a species native to coastal regions of eastern Mexico, Texas,

Louisiana, and coastal Mississippi—have become established. Greenhouse Frogs were documented in Alabama by Carey (1982), who found them at a junior high school in Baldwin County and speculated that they arrived in potted plants imported from South Florida, where Greenhouse Frogs have been established for decades. Based on data from Florida, these frogs are capable of invading upland pine habitats of the Coastal Plain, becoming abundant in Gopher Tortoise (*Gopherus polyphemus*) burrows (Meshaka 2011). Therefore, we consider this species to be established in the state. Rio Grande Chirping Frogs were first documented by Raymond and Tamara McConnell based on a series of specimens discovered in a heavily disturbed site in Mobile County (McConnell et al. 2015). Finally, Gulf Coast Toads were first detected in Alabama in Montgomery County by Birkhead et al. (2017). Because subsequent searches for these toads found them to be concentrated in new housing developments where transport in landscape vegetation seems likely, we consider this species to be non-indigenous. Had it arrived via range expansion from Mississippi, we would have considered it to be native. We consider all three species to be established in the state and provide species accounts for each.

The path to establishment of non-indigenous species is not an easy one. Most aliens die sooner or later and leave no trace of their former presence because insufficient founder populations are created to assure persistence. Nevertheless, records of species that failed to become established are important to the development of predictive models for understanding which species will invade and why. We know of three species of frogs that have been captured as free-ranging individuals in the state, but for which established populations have not been documented. One, White's Treefrog (*Litoria caerulea*), was observed in Homewood, Alabama. This native of Australia is sold frequently in the pet trade and likely escaped or was purposefully released into an area where that individual continued to survive. The second species is the Pacific Treefrog (*Pseudacris regilla*), a lineage from the western United States that was purposefully introduced to a yard in a rural area of Lee County in the early 1990s. Several individuals of both sexes were released, and calling males were heard for several years after the release. We have no evidence that calling males remain and, therefore, do not consider the species to be part of Alabama's fauna and provide no species account for it. However, this species is a member of a genus that is native to Alabama and, therefore, additional efforts need to be made to determine whether the founders established a viable population. Finally, an adult Cuban Treefrog (*Osteopilus septentrionalis*) was detected in a residential area of Alabama in 2012 (AUM 40229). This invader from the Caribbean has become established throughout much of Florida and seems destined to become established along the Gulf Coast of Alabama. The individual discovered in Auburn likely was transported to the area in potted plants originating from nurseries in Florida.

Taxonomic Changes and Problems

Our taxonomic list includes 32 named lineages (species, subspecies, or genetic clades) as being native to the state, a number that is second in richness only to Texas among US states. Our list represents a modest increase from the 30 lineages listed in Mount (1975). However, this number masks additional taxonomic changes indicated here. All taxa on our list have valid scientific names, but in the species account for American Bullfrogs (*Lithobates catesbeianus*) we indicate an additional unnamed taxon.

Based on Austin and Zamudio (2008), we reduce the two subspecies of Green Frogs (*L. clamitans*) recognized by Mount (1975) to a single variable species. Additionally, we eliminate the Northern Leopard Frog (*L. pipiens*) as being a possible member of the Alabama frog fauna. Mount (1975) interpreted leopard frogs from the northern portion of Alabama to show the influence of Northern Leopard Frogs, but recent interpretations place that taxon no farther south than the state of Kentucky.

We add two native species to Alabama's anuran fauna, both listed by Mount (1975) as approaching the state but not documented from within it. The Pine Barrens Treefrog (*Dryophytes andersonii*) is endemic to the pitcher plant bogs of Covington, Escambia, and Geneva Counties and was first documented for the state by Moler (1981). The Northern Crawfish Frog (*L. areolatus circulosus*) has been documented in Sumter County (Holt 2015). Finally, we use Newman and Rissler (2011) to document the presence of Florida Leopard Frogs (*L. sphenocephalus sphenocephalus*) and Coastal Plains Leopard Frogs (*L. s. utricularius*) in Alabama.

CLIMATE OF ALABAMA

Because of its location, with a southern border along the Gulf Coast and a northern border along the southern extent of the Appalachian

Mountains, the climate of Alabama is classified as humid subtropical (McKnight and Hess 2000). This climate is characterized by mild winters and hot, humid summers. Mean temperatures are warmer and more constant in the southern portion of the state than the northern portion, both because of lower southern topography and a stronger influence of Gulf breezes in the south. Rainfall is distributed throughout the year because of cold, dry polar fronts moving against warm, moist coastal air during autumn and winter, yielding intense thunderstorms, and moist warm Gulf air moving north during spring and summer, rising over terrestrial areas, and generating afternoon rains. Spring thunderstorms often occur with little rain, providing an ignition source for frequent low-intensity fires that maintain open-canopied pine forests with a lush herbaceous understory across much of the Coastal Plain. This feature maintains wetlands embedded in the region. Rainfall is slightly higher along the coast of Alabama because of the increased moisture content of the air associated with the Gulf of Mexico. Measurable snowfall is exceedingly rare in the southern half of the state, and annual totals of more than 6 in (150 mm) are seldom recorded even for the northernmost stations.

These typical patterns of weather are broken by annual occurrences of violent weather associated with tornadoes, mostly during spring, and hurricanes during late summer and fall. These storms can cause periods of intense rains that saturate soils and flood extensive areas. Such occurrences cause some terrestrial anurans to move to upland areas to avoid advancing waters and others to take advantage of opportunities to migrate from one wetland to another. These storms also kill trees by tipping them up from the roots, snapping them off at the trunk, or severely stressing them from saltwater storm surge. Each of these add fallen logs and leaves that accumulate on the floor of Alabama's forested habitats, thus providing sites used as refuges for frogs and toads during extreme temperatures, as well as providing wetlands used for reproduction. Frequent fire is vital to maintaining these wetlands. Intense breeding aggregations of anurans can occur during these storms, and these animals can become a rich food resource for other predatory vertebrates. In short, Alabama's climate is mild and moist enough, the geography diverse enough, and disturbances associated with storms frequent enough to support many species of frogs and toads.

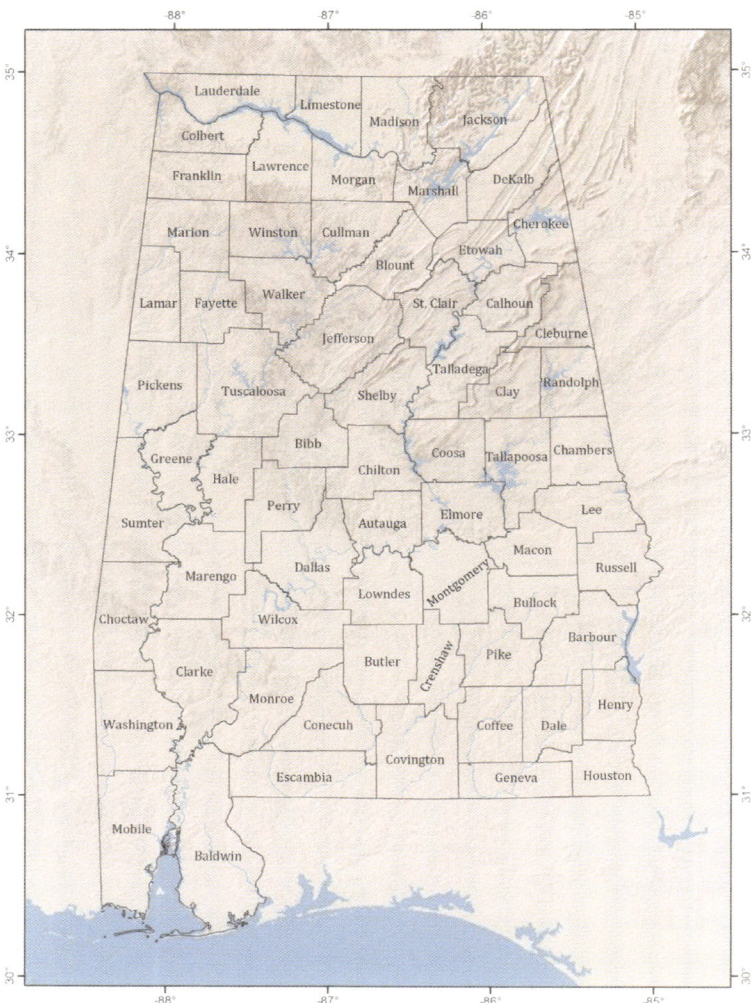

Counties of Alabama.

ALABAMA GEOGRAPHY

Alabama covers 52,419 mi^2 (135,765 km^2) of land and water in the southeastern US, essentially all of which is habitable by anurans. Although frogs and toads do not care about political boundaries, humans who study them do. For this reason, the state is divided into 67 counties that we will refer to because they are important frames

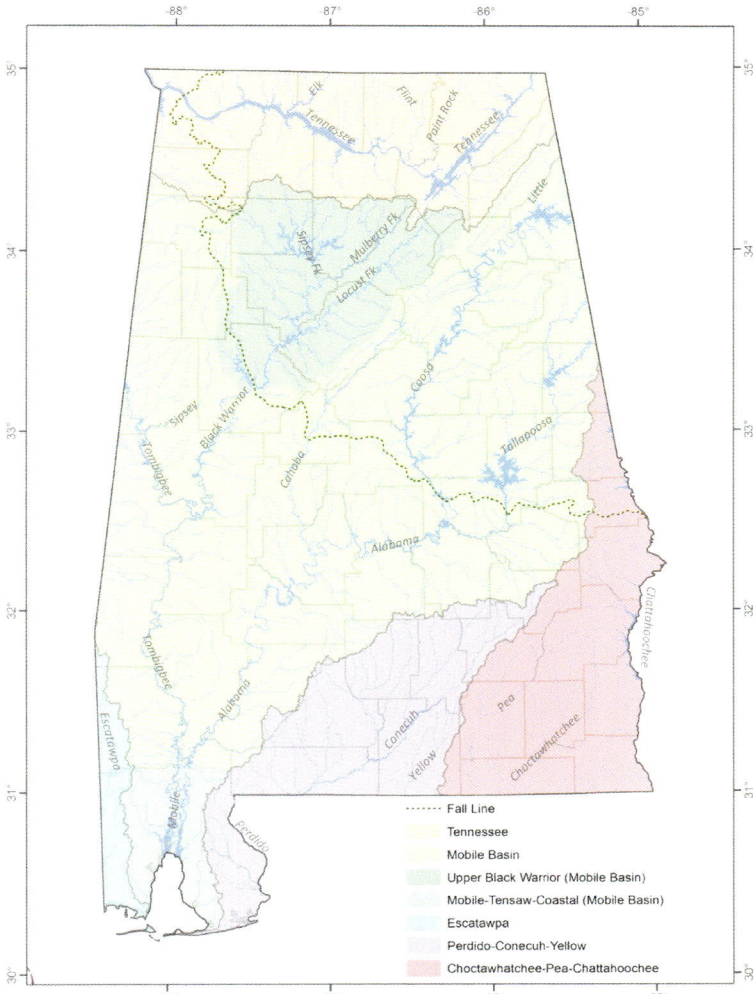

Major rivers of Alabama.

of reference for those reading this book. The distribution of anurans across the state are affected by geographic features that cross county boundaries, linking groups of counties that share river drainages or geological formations. Here, we summarize major geographic features of the state and discuss their effects on Alabama's anuran fauna.

Wetlands

One remarkable feature of Alabama's geography is the diversity of the rivers that drain its surface. Eight river systems are found in the state, all of which open to the Gulf of Mexico. Three of these are major rivers wide enough to present a challenge to most terrestrial organisms attempting to cross them. The Tennessee River enters the northeast corner of the state, flows east to west, and exits the northwest corner. This river drains the northern one-eighth of the state and joins the Mississippi River system, exiting into the Gulf of Mexico in Louisiana. Two additional major river basins are present in the state. One, the Chattahoochee, drains 6 percent of the state and forms the southern half of the eastern border; the other, the Mobile, drains most of the state (64 percent), entering the Gulf of Mexico near the southwestern corner. These are ancient rivers that have remained in close association with upland regions of the southern Appalachians. However, their connections to the Gulf of Mexico have lengthened or shortened because of marine inundation or subsidence. These physical processes altered terrestrial lowlands during the approximately 300 million years during which amphibians have occupied this region.

Black Belt pasture pond, Sumter County.

Five smaller rivers are also found in the state. The Escatawpa originates in southwestern Alabama, draining the western portions of Mobile and Washington Counties, about 3 percent of Alabama's surface area, but exits the state into Mississippi, where it opens to the Gulf of Mexico in Pascagoula Bay. The Perdido River drains Escambia and Baldwin Counties, opening into Perdido Bay. This short river drains only about 1 percent of Alabama. Two rivers, the Conecuh and Yellow, drain 8 percent of Alabama in the south-central portion of the state. These rivers leave the state, eventually entering Pensacola Bay in Florida. Receding shorelines likely connected these rivers over geological time. Finally, the Choctawhatchee and Pea Rivers drain about 6 percent of southeastern Alabama and unite before entering Florida and, eventually, Choctawhatchee Bay. These short rivers appear to share a common history with the Chattahoochee River.

Despite their obvious importance for all other major groups of Alabama's amphibians and reptiles, rivers appear to play a relatively minor role in explaining the distribution of frogs and toads. The one

Lower Coastal Plain limestone sinkhole pond, Covington County.

Roadside ditch,
Covington County.

exception appears to be the Gopher Frog (*Lithobates capito*) and Dusky
Gopher Frog (*L. sevosus*), which appear to be separated by the Mobile
River drainage.

Instead of rivers, isolated wetlands associated with stagnant or
slowly moving water most strongly affect where frogs and toads occur.
Most species require fish-free pools to reproduce, because when predatory fish are present, their ability to consume frog eggs and larvae
is strong enough to eliminate many species of anurans. Those species that have adapted to occupy permanent ponds and lakes, where
predatory fish are abundant, all develop tadpoles that are distasteful
to fish, allowing the tadpoles to survive and grow. For this reason, species such as American Bullfrogs and Green Treefrogs (*Dryophytes cinerea*) are the dominant vertebrates of ponds and lakes. Other frogs
seek pools of water that remain fish-free because they dry every year
or once every few years, preventing fish from becoming established.
For example, sinkhole ponds in limestone areas are prime breeding
sites for Barking Treefrogs (*D. gratiosa*) and Gopher Frogs. Still other
frogs seek temporary puddles of water created by low-lying topography
or small soil disturbances in the natural landscape and human activities in modern landscapes. Roadside ditches are a primary example of
such wetlands, which are frequented by Cricket Frogs (*Acris*), Chorus

Pitcher plant bog, Covington County.

Frogs (*Pseudacris*), North American Narrow-mouthed Toads (*Gastrophryne*), and North American Toads (*Anaxyrus*). Alabama's anurans do occupy wetlands with moving waters, but these tend to be specialized wetlands, such as pitcher plant bogs occupied by Pine Barrens Treefrogs or vegetated borders of small streams where predatory fish are rare or where vegetation prevents them from finding all eggs and tadpoles.

Important Geographic Units

From the perspective of understanding biotic diversity, Alabama is divisible into 10 meaningful physiographic units, which Mount (1975) termed herpetofaunal regions. These units fall into two natural groupings, Coastal Plain units of the southern part of the state that were created by ancient seashores, and upland units that were formed by the southern end of the Appalachian Mountains. These two groupings have a distinct boundary, the Fall Line, where streams change from rocky, fast-flowing waterways of upland units to the sandy or muddy, slow-moving waterways of the Coastal Plain.

Major physiographic regions of Alabama.

THE COASTAL PLAIN

The Coastal Plain in Alabama is distinctly belted and is physiographically more variable than it is in either Georgia or Mississippi. Topographically, it varies from flat to almost montane. The soils vary from acid sands and sandy loams, the dominant soils of the southern and northern belts, to heavy, calcareous, alkaline types, soils that dominate the regions of the central belts. The sandy soils are covered with pine forests, especially Longleaf Pines (*Pinus palustris*). Those soils that are not primarily sandy allow for the formation of steep ravines that support dense hardwood forests and include upland areas with heavy soils that are covered by grasslands lacking a pine overstory.

Rocks and rock outcrops occur in some portions of the Coastal Plain, but seldom to the extent that they do in provinces above the Fall Line. Most streams of the Coastal Plain are sluggish and have sand, silt, or gravel bottoms, but some flow over bedrock. Several of the streams have broad, low floodplains with sloughs and oxbows. Swampy habitats are common, many of them having been created by beaver dams.

From a herpetofaunal perspective, the Coastal Plain contains most of the richness of frogs and toads in Alabama. A total of 12 lineages is restricted to this region, representing 38 percent of the state's native fauna. Two of these, the Southern Toad (*Anaxyrus terrestris*) and Bird-voiced Treefrog (*Dryophytes avivoca*), are widespread across the Coastal Plain, tending not to be found outside of this geographic region. The rest are restricted to components of the Coastal Plain described below or have unique patterns of presence and absence across these components.

Lower Coastal Plain

The Lower Coastal Plain extends completely across the state and includes the southernmost tier of counties and the lower portions of some of those in the second tier. Relief varies from flat to gently rolling. This region encompasses about 20 percent of Alabama's total surface area, and soils are sands, sandy loams, and sandy clays, with occasional gravelly phases. The Lower Coastal Plain is delimited to the north by its boundary with the Red Hills region. The transition is rather abrupt west of the Conecuh River. Eastward the distinction is not well defined, and in eastern Alabama the Red Hills region tends to lose its biogeographic and physiographic integrity relative to that of the Lower Coastal Plain.

The dominant forest communities originally occurring on the upland sites are fire-maintained ones, such as sandhills dominated by Longleaf Pine and Turkey Oak (*Quercus laevis*) and pine flatwoods, although other types are also represented. Intensive forestry is practiced in much of the Lower Coastal Plain, with a recent trend toward replacing existing off-site Loblolly Pine (*Pinus taeda*) and Slash Pine

(*Pinus elliottii*) stands with Longleaf Pine. The intensive mechanical and chemical site preparation that often precedes replanting, however, tends to eliminate many native plant species, including wiregrasses (*Aristida beyrichiana* and *Sporobolus junceus*) and bluestem (*Andropogon* spp.), the dominant herbs in many natural communities within the Lower Coastal Plain. Because these grasses require prescribed fires to persist in modern times, extensive intact areas of natural communities have become quite scarce.

Streams of the Lower Coastal Plain are highly variable in character. The major rivers flow generally southward and along some stretches have broad, low floodplains. The smaller streams frequently are sand bottomed and tea colored. Some stretches of the latter have typical floodplain forest development along them, while others have high banks forested with pine or oaks. Swampy places along the smaller, slow-moving streams may be dominated by cypress (*Taxodium ascendens* and *T. distichum*) or thickets of titi (*Cliftonia monophylla* and

Lower Coastal Plain Longleaf Pine (*Pinus palustris*) flatwoods, Covington County.

Cyrilla racemiflora). Most natural ponds and lakes are of sinkhole origin, with clear, shallow water and abundant aquatic vegetation. The largest of these is Lake Jackson, which is situated on the Alabama-Florida boundary at Florala, Covington County. Additionally, because the topography is so limited, causing rivers to meander, oxbow ponds are common lentic habitats.

The dominant feature of the coast of Alabama is Mobile Bay. To the southeast, the bay is delimited by Fort Morgan peninsula, a narrow strip of land extending westward from the body of the mainland for about 18 mi (29 km). Fort Morgan peninsula, as well as the remainder of the coast of Baldwin County, is dominated by dune communities and xeric pine flatwoods, with some salt marsh habitat on the leeward side. West of Mobile Bay, in Mobile County, the coast of the mainland is low with extensive salt marsh communities. Several islands lie offshore of Mobile County, the largest of which is Dauphin Island. Dauphin Island is about 5 mi (8 km) long and about 1.5 mi (2.5 km) wide at the

Above: Blackwater stream, Escambia County.

Right: Cypress swamp, Clarke County.

widest point. The island has beach and dune habitat, as well as pine flatwoods and salt marshes. There is one small freshwater lake. Virtually unchecked coastal development in the form of condominiums, beach houses, commercial districts, and roads has significantly altered much of Alabama's coastline to the detriment of many anuran taxa.

Oxbow pond, Macon County.

The Pine Barrens Treefrog, Southern Chorus Frog (*Pseudacris nigrita*), Little Grass Frog (*P. ocularis*), Gopher Frog, Dusky Gopher Frog, Pig Frog (*Lithobates grylio*), River Frog (*L. heckscheri*), and Florida Leopard Frog have geographic distributions largely restricted to the Lower Coastal Plain, as does an unnamed lineage currently included within American Bullfrogs. Several of these species have spotty distributions, require rare habitats, and/or maintain low population densities, making them difficult to detect. Because of these features, the Dusky Gopher Frog is protected under the ESA, and the Gopher Frog is proposed for listing.

Lower Coastal
Plain beach
dune vegetation,
Baldwin County.

Red Hills

The Red Hills region is a belt of Eocene age (34–56 million years ago)
deposits that are 30–40 mi (48–64 km) in width and cut a semicir-
cular swath from the Mississippi border to the Georgia border. This
region encompasses about 10 percent of Alabama's total surface area,
much of which has deeply dissected topography, relatively fertile soils,
and frequent outcrops of siltstone, claystone, or limestone. In its western extent in Alabama (Choctaw, Clarke, Monroe, and Wilcox Counties) the Red Hills are almost montane in character, with rocky bluffs,
deep ravines, and clear rock-bottomed brooks. Here, the Red Hills
are abruptly differentiated from the Lower Coastal Plain to the south
and the Black Belt to the north. Some of the ridge tops rise as high as
300 ft (91.4 m) above the floors of the intervening valleys. The ridge
tops and upper ravine slopes tend to support communities with mix-
tures of pine and hardwood, while the coves and lower slopes often
have luxuriant hardwood forest stands, with oaks, hickories, beeches,
and magnolias predominating. At its eastern extent in Alabama, the
Red Hills topography is characterized by rolling hills that grade less
abruptly and interdigitate with the Lower Coastal Plain and Black Belt.

Some elements of the biota of the Red Hills are those usually associated with regions above the Fall Line and others are found in no
other Coastal Plain province. For example, the steep forested slopes

Red Hills ravine slope, Butler County.

of this region include disjunct populations of trees and shrubs typically found in the Appalachian Plateaus and Piedmont regions farther north. This feature is reflected in the anuran fauna only in the presence of disjunct populations of the Pickerel Frog (*Lithobates palustris*) in the Red Hills. Much of the Red Hills is in private ownership, but recent purchases of land by Alabama's Forever Wild Program have expanded public holdings on which conservation actions can be planned.

Black Belt

The Black Belt is roughly crescent shaped and extends almost continuously from western Tennessee southward through Mississippi and generally east-southeastward across central Alabama. It is about 35–40 mi (56–64 km) wide at its widest point in extreme western Alabama. This region covers 8 percent of the total surface area of the state. In Alabama, the western portion of this formation is distinctive and easy to differentiate from the Fall Line Hills to the north and the Red Hills to the south but becomes poorly defined at its eastern terminus well west of the Chattahoochee River in Russell County. The Alabama River forms the northern boundary of the Black Belt for the eastern half of the state and the Cahaba, Black Warrior, and Tombigbee Rivers dissect the Black Belt in the western half. These rivers are of low gradient and have extensive floodplains originally forested with

Black Belt
prairie remnant,
Autauga County.

hardwoods, cypress, or extensive bamboo patches that have largely been converted to agriculture. Ponds along these rivers are common, forming breeding habitat for anurans, and the banks are often steep and covered with lush vegetation.

The Black Belt is characterized by a predominance of heavy calcareous soils of Cretaceous origin (65–145 million years ago), creating gently rolling topography on which prairie vegetation occurs. Barren outcrops of Selma Chalk and forests of hardwood and mixed pine and hardwoods occurred naturally along riparian zones (Schotz and Barbour 2009). Fire maintained the tall grasses that carpeted natural prairie vegetation, and the area was rich in flowering plants. However, a near-complete conversion of this region to agriculture reduced the ancestral biota to very small parcels of land, most of which have become heavily encroached by trees and shrubs that invaded due to a lack of fire.

Scattered throughout the Black Belt region are inclusions of acid soil, some of which are sandy in texture and, therefore, seem appropriate for burrowing anurans. Several species commonly associated with light, friable soils in the Coastal Plain, however, appear to be scarce or absent from much if not all of the Black Belt. These include Oak Toads

Fall Line Hills fire-managed pine woodland, Macon County.

(*Anaxyrus quercicus*), Pine Woods Treefrogs (*Dryophytes femoralis*), Barking Treefrogs (*D. gratiosa*), Cope's Gray Treefrogs (*D. chrysoscelis*), and Ornate Chorus Frogs (*Pseudacris ornata*). The lack of extensive upland forests may also contribute to the scarcity of some of these forms. Interestingly, the one anuran species restricted to this region, the Northern Crawfish Frog (*Lithobates areolatus circulosus*), uses burrows created by crayfish as refuges.

Fall Line Hills

The Fall Line Hills, called the Upper Coastal Plain by some authorities and Central Pine Belt by others, lies between the Black Belt and the Fall Line. The region has a crescent shape that is widest along Alabama's western border with Mississippi and narrows to about 5 mi (8 km) along the Georgia border in Lee and Russell Counties. Topographically, it varies from moderately hilly in the west to gently rolling in the east, and in surface area it represents about 20 percent of the state.

The soils are Cretaceous in origin, mostly well drained, and vary from clay to sand. Gravelly phases are common. The sandy, well-drained sites often support communities dominated by Longleaf Pine and Turkey Oak, the best developed and most extensive of these being

Beaver pond, Marshall County.

in Autauga and Russell Counties. These communities are generally like those that occur on dry, sandy sites in the Lower Coastal Plain, and some support populations of Southern Toads. The Gopher Frog apparently once ranged over much of the Fall Line Hills region but appears to have been extirpated in recent times.

Creeks and rivers in this region typically have broad floodplains where oxbow and beaver ponds are common. In the Fall Line Hills in northwestern Alabama, there are several areas where streams have cut through the Coastal Plain sediments into the underlying sandstone, producing habitats like many of the gorges in the Appalachian Plateaus region to the east. Such habitats are exemplified by The Dismals in Franklin County. Physiographically, similar sites can be found at several other places in the Fall Line Hills in Franklin County, as well as in Colbert, Marion, Fayette, and Tuscaloosa Counties. Although of no known importance to explaining anuran distributions in the state, these are areas of importance in explaining distribution patterns of other plants and animals.

The Upland Regions

Piedmont stream bank, Lee County.

The upland regions, considered collectively to encompass the faunal provinces above the Fall Line, include the Piedmont, Talladega Upland, Ridge and Valley, Appalachian Plateaus, Tennessee Valley, and Highland Rim. The ranges of three anuran lineages, or approximately 9 percent of the Alabama fauna, are limited essentially to one or more of these regions. Two of these, the Eastern American Toad (*Anaxyrus americanus americanus*) and Mountain Chorus Frog (*Pseudacris brachyphona*) are widespread throughout the uplands.

Piedmont

The Piedmont extends into Alabama from the east and occupies a triangular area representing about 10 percent of the state's surface area in the eastern central portion. Along its lower margin, where it contacts the Coastal Plain, the transition is relatively abrupt, biotically and physiographically. To the north, the Piedmont borders the Talladega Upland. The transition there is not abrupt, and the northern

Talladega
Uplands near
Mount Cheaha,
Clay County.

portion of the Piedmont and the Talladega Upland have several features in common.

The Piedmont is hilly, for the most part, with clay soils that tend to be rocky and red. Much of the land is forested with Shortleaf Pines (*Pinus echinata*) and Loblolly Pines on the ridge tops and hardwoods along lower slopes and in the bottomlands. Granite outcrops are common throughout the region. Because the soil is so poor, extensive areas of agricultural lands have reverted to forested tracts. The Piedmont is drained by creeks that have rocky slopes, typically covered by fallen logs and with hardwood and pine trees forming the canopy. Stream channels have rock bottoms that guide swift-flowing, well-aerated water.

The anuran fauna of the Piedmont contains no anuran lineage endemic to it but has four lineages that generally reach their northern boundary in this region; these are the Southern Cricket Frog (*Acris gryllus*), Barking Treefrog, Pine Woods Treefrog, and Squirrel Treefrog (*D. squirella*).

Talladega Upland

Above the Piedmont is a series of ridges, extending approximately 100 mi (160 km) southwestward that we term Talladega Upland, following Griffith et al. (2001), but that was designated Blue Ridge by Mount (1975). The region covers about 1 percent of the surface of Alabama and has been considered by most recent accounts to be a subunit of the Piedmont. We treat it separately because of its apparent role in limiting distributions of some anuran species.

The Talladega Upland gives way abruptly to the Coosa Valley, the main body of which lies to the northeast. As previously noted, the transition to Piedmont is rather gradual. The highest point in Alabama, Mount Cheaha (elevation 2,407 feet [730 m]), is a component of the Talladega Upland, as are a few other peaks in the region that exceed 2,000 ft (600 m) in elevation.

The soil of the Talladega Upland tends to be rocky and friable, with some sandy phases. A great majority of the region is devoted to forestry, and a variety of forest habitat types are represented. Longleaf Pine is one of the most abundant trees on drier sites. Small streams arising within the region are cool and clear and typically retain hardwood cove forests along the slopes leading to the stream headwaters, and hardwoods dominate the riparian zones. One species, the Wood Frog (*Lithobates sylvaticus*), is largely restricted to higher elevation sites within the Talladega Upland.

Ridge and Valley

The Ridge and Valley region lies between the Talladega Upland and the Appalachian Plateaus. It extends southwestward from DeKalb County, near the northeast corner of Alabama, to the Fall Line, and covers about 9 percent of the surface of the state. The region is considered here to consist of the area between the Coosa and Cahaba Valleys, and the uplands arising within these valleys. The ridge portions emerge from non-erodible rock formations that create steep slopes of remarkably straight orientation caused by the northeast-to-southwest folding of the parent materials in the Ordovician Period (490 million years ago). The steep slopes drop to narrow valleys created by rivers that removed erodible parent materials to create the current geography. The rocks at the lower elevations are mostly limestone and shale, with sandstone and chert at higher elevations. The Coosa Valley occupies the largest area. The most prominent ridge is Double Oak Mountain, most of which lies in Shelby County, near the region's southwestern terminus.

The complex soils that comprise the Ridge and Valley formation allow water to accumulate in aquifers because of the porosity of the carbonate and sandstone layers. Underground conduits of increasing size eventually allow this water to reach the surface as springs that are common in this region, including some of the largest springs in the state. Additionally, sinkholes and caves develop through this process, all of which create habitat exploited by Alabama's anurans. No anuran species is restricted to this region.

Appalachian Plateaus

Lying immediately north of the Ridge and Valley are the Appalachian Plateaus, a region that covers about 15 percent of the surface of Alabama. Physiographically, these plateaus are considered subdivisions of the Cumberland Plateau, a geological formation that, like the Ridge and Valley, was formed by folding of continental materials during tectonic events of the Ordovician Period (490 million years ago). The plateaus that were created include Lookout Mountain and Sand Mountain south of the Tennessee River and extensive mountains in Jackson County north of the Tennessee River. Lookout Mountain originates in Tennessee, extends southwestward across the northwestern corner of Georgia and into Alabama, where it narrows and gradually loses its identity around Gadsden in Etowah County. Sand Mountain lies to the west and northwest, across the valley of Big Wills Creek. At the Alabama-Georgia boundary, Sand Mountain is about 15 mi (24 km) wide. Southwestward it expands to occupy much of the north-central portion of

Alabama north of Birmingham. The integrity of Sand Mountain as a plateau is maintained westward to within about 20 mi (32 km) of the Fall Line before it breaks up into an area of deeply dissected terrain in Lawrence and Winston Counties. Much of this rugged terrain is included within the Bankhead National Forest.

Lying north of Sand Mountain, from Morgan County westward, is Little Mountain. This narrow, somewhat irregular ridge has biogeographical characteristics like those of Sand Mountain; it is separated from the latter by Moulton Valley, a narrow, low-lying intrusion of the Tennessee Valley.

The soils of the Appalachian Plateaus are mostly sandy loams, although dry soils are not uncommon in the southern portion and at some of the lower elevations in the north. Rocky phases are found at many sites. Throughout most of the region sandstone is the dominant exposed rock. Shale and limestone as well as chert are common in the southern portion.

The gently rolling tops of the plateaus, especially portions of Sand Mountain, are often intensively farmed, but the edges and sides have much of the natural habitat remaining. The streams draining protected watersheds are clear, often emerge from springs with rock and sand bottoms, and frequently drop from plateau overhangs to valleys below creating waterfalls. Similarly, caves are common in this region, and streams frequently emerge from their mouths or fall into them via sinkholes. Sag ponds, depressional wetlands that develop along slip-strike faults, can be found along the flat tops of the Appalachian Plateaus, providing breeding sites for many anurans. No lineage of anuran is confined to the Appalachian Plateaus, but the Pickerel Frog is a characteristic component of its steep slopes and caves.

Above: Cave entrance, Appalachian Plateau, Jackson County.

Right: Ephemeral sag pond, Madison County.

Tennessee Valley

The main body of the Tennessee Valley is a broad expanse of relatively level, fertile land that lies along the Tennessee River and covers about 4 percent of the surface area of Alabama. To the east and south, where the valley meets the Appalachian Plateaus, there are bluffs and steep slopes, the lower reaches of which have extensive limestone outcrops. The soil of the valley is mostly red clay of limestone origin, and the few remaining forests are composed mostly of hardwoods and scattered patches of Eastern Red Cedar (*Juniperus virginiana*).

The Tennessee River is impounded throughout its course in Alabama and bears little resemblance to its former state. The shoals, which once contributed to a diverse aquatic fauna, are now all under deep water. No anuran species is restricted to the Tennessee Valley.

Tennessee River opposite Paint Rock River, Marshall County.

Highland Rim forested slope, Lauderdale County.

Highland Rim

North of the Tennessee Valley in northwestern Alabama is the Highland Rim, a region termed the Chert Belt by Mount (1975). Physiographically the Highland Rim is the southernmost subdivision of a vast province whose components are termed collectively the Interior Low Plateaus. In Alabama, the Highland Rim covers about 3 percent of the total land area, including much of Lauderdale and Limestone Counties, and portions of Madison, Colbert, and Lawrence Counties.

The Highland Rim is a moderately elevated region, with topography varying from hilly to nearly flat. The region lies entirely within the Tennessee River drainage area and the greatest relief is typically found near the streams. The soils are mostly heavy and fertile and once supported extensive hardwood forests. In the current landscape the areas that remain forested are mostly in Lauderdale County.

The Mountain Chorus Frog has a disjunct population in the Highland Rim. Southern Cricket Frogs appear to reach their northern limit in Alabama in the Highland Rim, and the Eastern Spadefoot (*Scaphiopus holbrookii*) and Eastern Narrow-mouthed Toad (*Gastrophryne carolinensis*) appear to be absent from this formation despite being widespread in formations around it.

Species Accounts

The remainder of this book describes anurans as a major radiation of tetrapods (land vertebrates), presents each family found in Alabama, and then summarizes information about each of the state's species within each family. Presentation of each family, genus, species, and subspecies is in the order of appearance within the keys provided. Important genetic variation supported by published analyses is discussed within each species account. Each species or subspecies account has distinct sections that we summarize below.

KEYS

Keys are tools designed to aid in identification of organisms. These tools present paired descriptions, one of which will conform to an individual organism of interest and the other description will not. At the end of each consistent description is a number indicating the next couplet to be considered. This process of making dichotomous choices is followed until a final description identifying the organism of interest is reached. We include taxonomic keys for the frogs and toads of Alabama and take the unusual step of dispersing these keys throughout

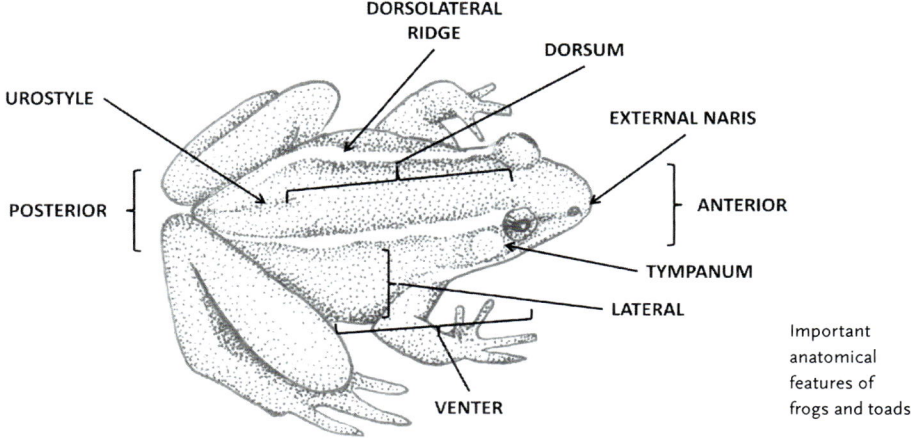

Important anatomical features of frogs and toads.

the accounts rather than including a single key. We do this to place information close to sections of text for which the keys are most useful. Keys to the families of anurans appear at the end of the description of the group Anura. When they are necessary, keys to genera are placed at the end of the description of each family, keys to species appear at the end of descriptions of each genus, and keys to subspecies are placed at the end of the Taxonomy section of the appropriate species account.

NAMES

The generic, specific, subspecific, and common names applied are, in most cases, those listed in Crother et al. (2017). In cases for which genetic data document potential novel lineages, we use common names generated by the phylogeographic studies.

PHOTOGRAPHS

We have benefited from the talents of many photographers. Where possible, we have selected pictures that show key features rather than those that have the best background or artistic composition. When the location of the specimen photographed is known, we identify it.

DESCRIPTIONS

Our descriptions are intended to provide sufficient information to enable the reader to distinguish a particular taxon from all others occurring within the state. Each description is based on a composite of specimens representing variation within Alabama and surrounding states. But, because nature is variable, it should be kept in mind that occasional individuals belonging to the described taxon will not conform to the descriptions presented here.

ALABAMA DISTRIBUTION

In addition to a general statement describing the distribution of each taxon occurring within Alabama, a map is included. For each map, we include diagonal hatching intended to delineate the presumed range at the state level. As these data are incomplete for most taxa, our estimated ranges should not be expected to be entirely accurate. Maps for species thought to occur statewide lack hatching. Known occurrences are presented as red dots on a herpetofaunal regions base map, and readers may use these, along with the discussion in the text, as further clues to overall distribution. These red dots depict specific locations for specimens the authors have examined; photo-vouchered specimens submitted to the Alabama Herp Atlas Project (a web-based program of citizen-scientist records stored at AUM); occurrences documented in the databases of the Alabama Natural Heritage Program and/or ADCNR State Lands Division, Natural Heritage Section; or literature records believed valid. Rebecca Bearden, Brian Folt, Ashley Peters, Eric Soehren of the State Lands Division, and Michael Barbour and Jim Godwin of the Alabama Natural Heritage Program were particularly notable in providing important occurrence records. Each record was georeferenced and plotted to the greatest possible precision using ArcGIS® software produced by Esri. The state distribution maps also include insets showing the approximate distribution of each taxon within the continental United States. These are modified and customized from many previously published distribution maps, particularly those developed for Jensen (2008).

Habits

Here we provide information on seasonal patterns of reproduction, mating strategies of males and females, major diet items, and habitat preferences. In general, this section is designed to summarize where and when each species is likely to be active and what activities make the species detectable by humans. Additionally, we describe the timing and duration of each major growth stage in the life cycle of the species.

Conservation and Management

In this section we describe the current conservation status of each species in Alabama. Because conservation issues are likely to increase in the future, we summarize management activities that might imperil each species as well as those activities that are likely to enhance populations. For species that have conservation status, we provide information on key public properties that will play crucial roles in the long-term maintenance of Alabama's anuran fauna. In developing its Alabama's Wildlife Action Plan (Division of Wildlife and Freshwater Fisheries, ADCNR 2015), the State of Alabama used the findings from its 2012 Nongame Symposium, which assembled scientific experts to compile the best data on Alabama's wildlife, to identify the species most in need of conservation action. The Nongame Symposium's Amphibian and Reptile Subcommittee identified five anurans as being of immediate conservation need (Priority 1 or 2, on a scale of 1 to 5;

Shelton-Nix 2017), and we summarize the subcommittee's recommendation in each species account.

TAXONOMY

We accept the concept that species are lineages that are discovered through careful analysis of variation in the characteristics of organisms. These discoveries arise from the creation of phylogenetic trees based on character data. Under this species concept, any diagnosable terminal branch is sufficient to discover a new species. Additionally, we accept the concept that taxonomic groups at any level of classification should be monophyletic. Each member of such groups is more closely related to other group members than it is to any organism outside the group. In practice, ancestral species might survive through the branching process, generating some lineages that are not monophyletic. Such species present challenges for determining species boundaries, and the decisions that we make for the boundaries of Alabama's species undoubtedly will suffer from this challenge.

For anurans, color patterns, degree of webbing, presence and conformation of tubercles in adults, and shapes of tails in tadpoles are external features that traditionally have been used to diagnose species. To these traditional characters are added sequence data generated from the mitochondrial and nuclear genomes. These sequences have the advantage of allowing rapid development of data sets that are much larger than those based on morphology. The mitochondrial genome is inherited in offspring entirely from the female side of the family tree while the nuclear genome captures information about gene flow associated with both parents. For this reason, phylogenetic trees based on the mitochondrial genome are not guaranteed to be concordant with those based on the nuclear genome. Nevertheless, published data on the mitochondrial genome often are assumed to carry phylogenetic information in the absence of data to the contrary. These data are particularly important in phylogeographic studies, a field of biogeography that uses patterns of evolution within species that are discernable by analysis of molecular data. Such studies now are common for Alabama's anurans, and the intraspecific lineages generated by such studies will likely allow us to discover new species that were not evident from analysis of traditional morphological data. This creates an exciting environment for taxonomists because so many new species may be discovered, and each discovery reveals important information about how that diversity was generated (e.g., Soltis et al. 2006). It also creates a scary environment for authors of field guides, such as this one, because the taxonomy of the guide likely will be obsolete before it is published. Because of this we attempt to describe all lineages supported by character data that might indicate speciation events awaiting taxonomic recognition.

Mount (1975) was exemplary in recognizing important subspecific variation within Alabama. That these taxonomic distinctions were important is supported by the three subspecies (*Lithobates capito capito, L. c. sevosus, Pseudacris triseriata feriarum*) that have since been elevated to species status, largely based on the accumulation of molecular data. We retain subspecies that are based on characters that show apparently significant geographic discontinuities. Phylogeographic studies frequently find imperfect concordances between subspecies boundaries and boundaries of mitochondrial lineages, and this discordance frequently is used to argue against traditional subspecific boundaries (e.g., Austin and Zamudio 2008). In fact, this trend is so strong that large numbers of lineages have emerged that represent numbered or lettered clades on phylogenetic trees (e.g., Richter et al. 2014). Where possible, we attempt to align subspecific names with numbered or lettered clades and use such alignment to retain some subspecific categories. In other cases, mitochondrial data, especially in association with subspecific designations that appear to be based on clinal variation, are used to reject previous subspecific recognition. If no taxonomic names are available for clades discovered within phylogenetic studies, then we retain named, numbered, or lettered clades described in such studies.

Frogs and Toads

Anura

The group Anura comprises about 6,600 species and represents the dominant group of living amphibians as well as one of the most species-rich lineages of living tetrapods. Although diverse taxonomically, frogs and toads form a distinctive group whose members can hardly be mistaken for anything else. Adults lack a postanal tail, have exceptionally short trunks, and possess remarkably long hind limbs designed for jumping. Most have wide heads, with large, bulging eyes. As a rule, anurans can be placed into one of two categories: a group with relatively short hind legs used for hopping and a group with relatively long hind legs designed for leaping. Hoppers are designed to travel relatively short distances with each jump but can repeat this in succession over relatively long periods of time. Leapers are designed to travel great distances with each jump but can repeat this only a few times before becoming so physically exhausted as to require a period of recovery before being able to leap again. The hoppers frequently are referred to as toads and the leapers generally are referred to as frogs. Additionally, anurans referred to as toads frequently have thick, bumpy skin, while those referred to as frogs usually have smooth, thin skin.

The oldest fossil anuran dates to 230 million years before present (Rage and Rocek 1989), and the living descendants of the first frog now occupy all major continents. Although the basal clades tend to be found in temperate northern regions, multiple independent lineages of frogs have invaded tropical regions where they have experienced increased rates of speciation leading to the great species richness of modern forms (Wiens 2007).

The majority of frogs have males that produce a vocalization referred to as an advertisement call because this sound communicates the position of the calling male to other males in the chorus and to potential female mates. When in a chorus, males tend to concentrate on calling to the exclusion of other activities. Because it is energetically expensive to produce a call, calling males eventually must leave the chorus when their energy reserves are depleted. In some species, males compete for positions in the chorus through the physical defense of territories (e.g., most members of the family Ranidae), whereas others divide space evenly across the reproductive site but do not physically defend a calling site (e.g., most members of the family Hylidae). Because the dominant frequency of the call (call pitch) becomes lower in larger individuals of the same species, the call of the male also contains information about the size of the frog producing the call, a feature used by rival males to assess one another.

In most frogs, females use the qualities of the male advertisement call to select from among potential mates. Females generally select males with deeper voices (larger size) and longer, more vigorous calls because this is thought to indicate the fittest males (e.g., Welch et al. 1998). However, in a wide variety of species the female must move through non-calling satellite males. These males, which generally are smaller, position themselves near calling (larger) males and attempt to mate with the female attracted to the calling male (Leary et al. 2005).

Once a female has selected a mate, the male grasps the female from behind, whereupon the two are said to be in amplexus. In most frogs, the male grasps the female either in the groin (pelvic amplexus) or armpit (pectoral amplexus) and then hangs on while the female selects a site to deposit her eggs. This may occur in water or on land, and typically the eggs are fertilized externally by the amplectant male. Development may involve an aquatic larva (tadpole) that transforms into the adult body form or direct development of a fertilized egg into a fully transformed froglet. Similarly, parental care may be reduced, only involving placement of eggs into an aquatic nest, or may be complex, involving provisioning of the offspring by the female parent. Thus, anurans display a wide variety of reproductive modes. However, for native species in Alabama, all females deposit eggs in water. These are fertilized externally by the male, and then both parents leave the eggs to hatch on their own, develop as free-swimming tadpoles, and transform into the adult form. The time required to complete the aquatic phase of the life cycle varies from 12 days for the Eastern Spadefoot (*Scaphiopus holbrookii*) to as much as 2 years or more for the American Bullfrog (*Lithobates catesbeianus*) (Duellman and Trueb 1986).

Tadpoles have a variety of foraging modes. Most are herbivorous, possessing enlarged black keratinized beaks that scrape algae and microbes from aquatic substrates. These herbivorous forms typically have long, coiled intestines that can be seen through the opaque wall of the belly. Other tadpoles are planktivorous, pumping water into the mouth cavity to harvest microscopic organisms. Adults are entirely predaceous, mostly consuming arthropods, but with some species becoming large enough to eat vertebrates, especially other frogs.

Key to the Families of Anura of Alabama

1a Undersurface of hind foot with one or two hard spade-like tubercles on heel; tadpole with median anus; **go to 2**.

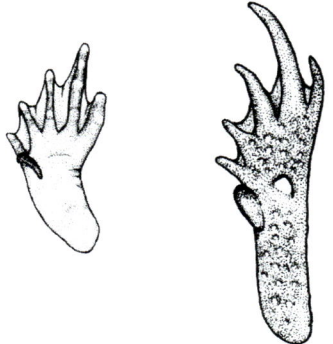

From left to right:

Ventral view of the hind foot of the Eastern Spadefoot (*Scaphiopus holbrookii*) showing a single spade-like tubercle.

Ventral view of hind foot of Fowler's Toad (*Anaxyrus fowleri*) showing two spade-like tubercles.

Ventral view of tadpole with median anus.

1b Undersurface of hind foot lacking hard spade-like tubercles on heel; tadpole with anus opening to the right (dextral); **go to 3**.

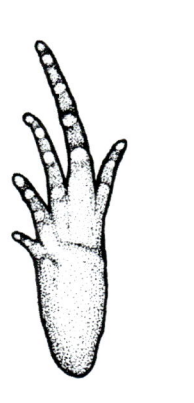

From left to right:

Ventral view of hind foot of Eastern Narrow-mouthed Toad (*Gastrophryne carolinensis*) showing no tubercle.

Ventral view of *Scaphiopus* tadpole with dextral (right-facing) anus.

2a Undersurface of hind foot with one hard spade-like tubercle (see illustrations with couplet 1); pupil of eye vertically elliptical; tadpole with complete oral disc.

Family Scaphiopodidae . . . page 45.

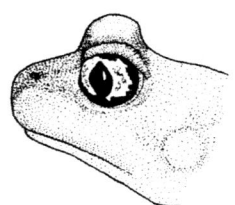 

From left to right:

Lateral view of head of *Scaphiopus* showing vertically elliptical pupil.

Ventral view of *Scaphiopus* tadpole mouth parts showing complete oral disc.

2b Undersurface of hind foot with two hard spade-like tubercles (see illustrations with couplet 1); pupil of eye horizontally elliptical; tadpole with interrupted oral disc.

Family Bufonidae . . . page 51.

From left to right:

Lateral view of head of *Anaxyrus* showing horizontally elliptical pupil.

Ventral view of *Anaxyrus* tadpole mouth parts showing interrupted oral disc.

3a Head with a transverse fold behind eyes (easiest to see in life); head and snout abruptly tapering to a point; tympanum absent; hind foot lacking webbing between toes; tadpole lacking keratinized teeth.

Family Microhylidae . . . page 79.

 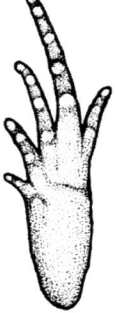 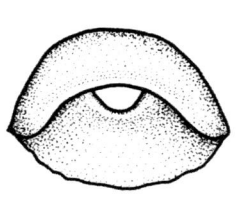

From left to right:

Dorsal view of Eastern Narrow-mouthed Toad (*Gastrophryne carolinensis*) showing fold of skin behind eyes and abruptly tapering snout.

Ventral view of hind foot of Eastern Narrow-mouthed Toad (*Gastrophryne carolinensis*) showing a lack of webbing between toes.

Ventral view of *Gastrophryne* tadpole showing jaws that lack keratinized teeth.

3b Head without a transverse fold; head and snout not abruptly tapering to a point; tympanum present; hind foot usually with at least some webbing between toes (see illustrations with couplets 4 and 5); tadpole with keratinized teeth (see illustrations with couplet 4); **go to 4.**

4a Belly skin smooth; hind foot with webbing extending more than halfway along fourth toe; tadpole with emarginated oral disc.

Family Ranidae . . . **page 85.**

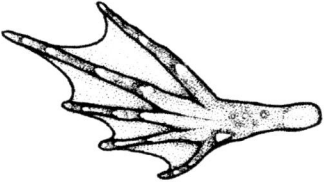 

From left to right:

Ventral view of hind foot of *Lithobates* showing webbing extending more than halfway along fourth toe.

Ventral view of mouth of *Lithobates* tadpole showing jaws with keratinized teeth and emarginated oral disc.

4b Belly skin granular; hind foot with webbing at most halfway along fourth toe (see illustrations with couplet 5); tadpole stage lacking or tadpole with entire oral disc; **go to 5.**

Ventral view of *Acris* tadpole mouth parts showing entire oral disc.

5a No intercalary cartilage on digits; webbing on hind foot a thin wisp at base of toes; direct development, tadpole stage lacking.

Family Eleutherodactylidae . . . **page 137.**

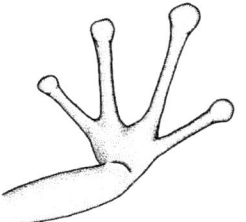 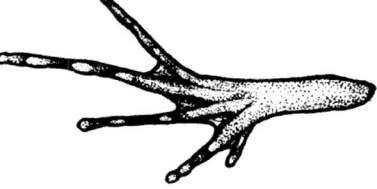

From left to right:

Ventrolateral view of hands of *Eleutherodactylus* lacking intercalary cartilages.

Ventral view of hind foot of *Eleutherodactylus* showing a thin wisp of webbing along fourth toe.

5b Intercalary cartilage present on digits; webbing on hind foot either absent or present to about halfway along fourth toe; tadpole with entire oral disc (see illustrations with couplet 4).

Family Hylidae . . . page 145.

From left to right:

Ventrolateral view of hands of *Dryophytes* showing presence of intercalary cartilage on each toe.

Ventral view of hind foot of Ornate Chorus Frog (*Pseudacris ornata*) showing webbing restricted to base of toes.

Ventral view of hind foot of Southern Cricket Frog (*Acris gryllus*).

Spadefoots

Family Scaphiopodidae

The family Scaphiopodidae is distinguished from other frog families in Alabama by the toad-like body (globose shape with short hind legs), vertically elliptical pupils (cat eyed), and the presence of a single, keratinized, spade-like tubercle on the heel of each hind foot. This family is found only in the US, Canada, and Mexico and includes species that live in extremely arid conditions and that are adapted for explosive reproduction. During these reproductive events males and females gather in great numbers at temporary wetlands for short periods of time, deposit eggs in great numbers, and leave. This is followed by the rapid growth and transformation of tadpoles. Between reproductive events Spadefoots spend long periods of time burrowed in loose soils. The family is sister to the families Megophryidae, Pelobatidae, and Pelodytidae, toad-like anurans found in Europe and Asia (Pyron et al. 2011). Therefore, like many other North American anurans, Spadefoots share a history with Europe and Asia. Two genera of this family, containing seven total species, occur within the United States. One genus is found in Alabama.

North American Spadefoots
Genus *Scaphiopus* (Holbrook, 1836)

This genus, which contains three species, occurs in the desert south-west of the United States and Mexico as well as regions with sandy soils in the eastern United States. It is the sister taxon to the genus *Spea*, the only other genus in the family. Members of *Scaphiopus* have the same general characteristics as the family. A single species occurs in Alabama.

Adult Eastern Spadefoot (*Scaphiopus holbrookii*), Covington County, AL.

Eastern Spadefoot
Scaphiopus holbrookii (Harlan, 1835)

DESCRIPTION Eastern Spadefoots are medium-sized anurans, attaining a maximum snout–vent length of about 2.7 in (70 mm). These animals have warty skin, hind feet that are completely webbed, pectoral glands, and eyes with vertically elliptical pupils. The dorsal ground color is olive to grayish brown, with a lyre-shaped dark figure framed by a pair of yellowish or greenish-yellow stripes extending from behind each eye to the region of the vent. No cranial crests are present, and the hind foot possesses a black, keratinized, spade-like tubercle on the inner surface of the inner toe. The tadpole has dorsal eyes, is bronzy brown in body color, and possesses clear tail fins.

Eastern Spadefoots can be confused with North American Toads, members of the genus *Anaxyrus*. However, North American Toads have horizontal pupils, two heel tubercles, and parotoid glands that are easily distinguished. Tadpoles of Eastern Spadefoots are like those

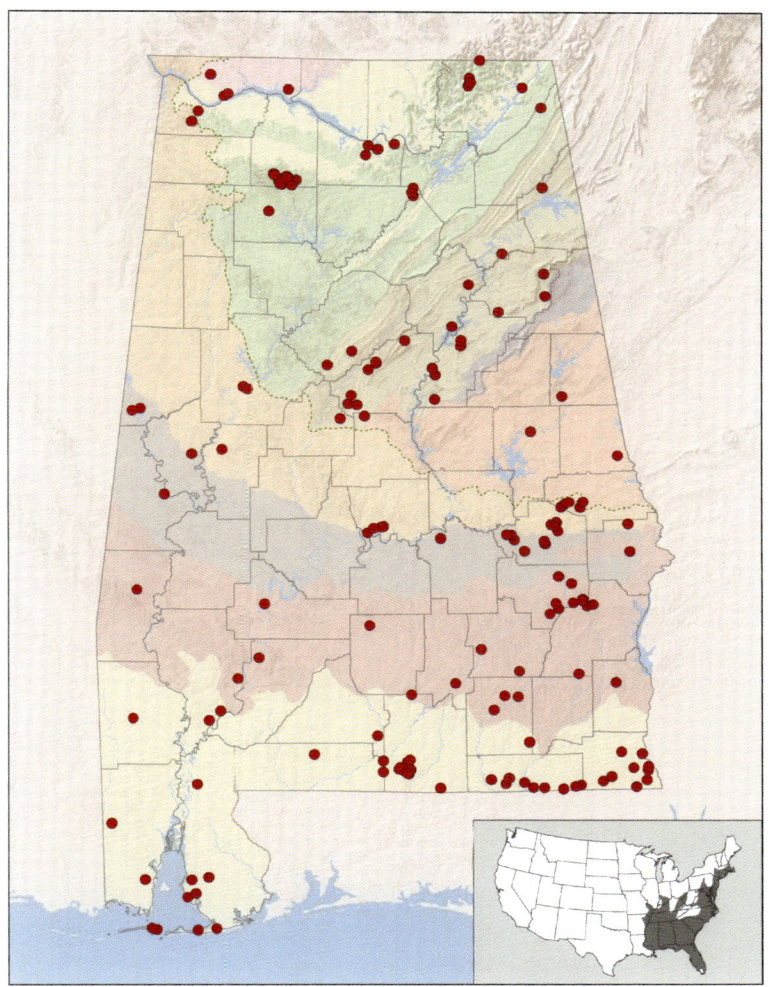

Distribution of the Eastern Spadefoot (*Scaphiopus holbrookii*). Solid dots indicate localities of specimens or photographs examined by the authors or ADCNR/Natural Heritage Program occurrence records believed to be valid. Inset map depicts approximate range in the United States.

of Southern Chorus Frogs (*Pseudacris nigrita*) and Mountain Chorus Frogs (*P. brachyphona*) in size and have clear tail fins, but the *Pseudacris* tadpoles have lateral eyes.

ALABAMA DISTRIBUTION Eastern Spadefoots are spotty in their distribution but are found throughout Alabama, especially where soils are sandy or loamy.

HABITS Eastern Spadefoots are a secretive, burrowing form, emerging from its refuge only at night or on heavily overcast days. Breeding is confined to temporary pools and ponds resulting from heavy rain. These sites lack predatory fishes and are surrounded by upland pine forests on sandy soils (Greenberg and Tanner 2004) or mixed pine-hardwood forests on loamy soils. Activity at breeding sites is triggered by intense warm rains (Timm et al. 2014). Such conditions typically occur during spring and summer, but breeding may take place in any month of the year. Adults migrate from their home ranges to breeding sites, traveling distances up to 0.25 mi (0.4 km) (Pearson 1955; Timm et al. 2014). The advertisement call of males is an explosive, low-pitched "waaaah" that finishes lower in pitch than it starts and is repeated at short intervals. Some have likened the call to that of young crows, and others have likened it to someone with gastrointestinal problems. The call is issued while the male sits or floats in water. A large chorus can be heard from 1 mi (1.5 km) or more and is one of nature's oddest sounds. Males exhibit pelvic amplexus, a reproductive behavior in which a male grasps a receptive female where her hind legs join her body. This is the only native anuran in Alabama to display this behavior. Females then lay stringy masses of 1,000–2,500 eggs that are fertilized externally by the male. Eggs hatch in 10–15 days, depending on temperature, and the tadpoles transform 14–25 days later. However, the wetland sites used by Eastern Spadefoots are extremely ephemeral and it is not uncommon for entire pools to dry before even a single tadpole has transformed. Growth of tadpoles of Eastern Spadefoots decreases as density increases, exacerbating the difficulty of generating successful reproduction. In such an environment, the fastest growing individuals are the ones that are most likely to metamorphose, and they do so at a small size. Those tadpoles that are adapted to survive in crowded conditions are next most likely to metamorphose, and they do so at a larger size (Semlitsch and Caldwell 1982). Juvenile recruitment is sporadic and typically is associated with seasons when breeding adults are numerous (Greenberg and Tanner 2005a). Leaf litter is a key resource used by juveniles as refugia as individuals migrate from the breeding site (Baughman and Todd 2007).

During the remainder of the active season, adults are nocturnal, with peak activity between 2100 and 0100 hours (Punzo 1992). Animals surface during an average of 29 nights a year, and most active time is spent feeding on beetles, termites, moths, and ants that are consumed within 3 ft (1 m) of the site where the Eastern Spadefoot burrows into the soil. Thus, adults spend far more time burrowed underground than they do on the surface during a year of activity, and the part of an individual's home range associated with feeding is exceptionally small. The species tends to aggregate in open, grassy, relatively shrub-free areas. Males grow faster than females and attain a larger size (Pearson 1955), an unusual feature for anurans. Adult life span is on the order of seven years (Greenberg and Tanner 2005a). Tadpoles are omnivores, consuming organic matter, phytoplankton, and microscopic organisms in the water column. These larvae also are social, schooling while foraging to create vortices in the water column that increase food material that can be processed.

CONSERVATION AND MANAGEMENT This species is common in areas with loose, sandy soils. For this reason, it receives no regulatory protection in Alabama. It requires fish-free wetlands for breeding purposes, but these can be remarkably ephemeral and, therefore, do not appear to be limiting. Plowing or root-raking activities are likely to kill adults in their burrows and might reduce population densities. Such management techniques should be avoided if retention of native biodiversity is a management goal. Eastern Spadefoots are incapable of burrowing in thick root mats produced by St. Augustine grass (*Stenotaphrum secundatum*; Jansen et al. 2001), a feature that is likely to be true in areas invaded by Cogongrass (*Imperata cylindrica*). Eastern Spadefoots also disappear from urban habitats where the wetland and upland habitats required by this species are altered (Delis et al. 1996). Mass mortality events associated with the increased prevalence of ranavirus are known for these frogs (Miller et al. 2011).

TAXONOMY We follow Frost et al. (2017) in considering this to be a single species with no subspecific variation. This is the sister species to Hurter's Spadefoot (*S. hurterii*) of eastern Texas, Louisiana, Arkansas, and Oklahoma east of the Mississippi River.

True Toads

Family Bufonidae

This large family, with 51 genera and about 560 species, is essentially cosmopolitan in distribution, except for its absence (excluding introduced species) in Australia, New Guinea, Madagascar, and Polynesia. These anurans are squat animals, with relatively short hind limbs designed for hopping, warty skin with well-developed toxin-producing glands, and a Bidder's organ, a remnant of the developing ovaries retained in mature males. All members of the family have aquatic eggs that are laid in two long strands, similar in appearance to strings of pearls, and deposited in temporary pools of water. Tadpoles occur in dense schools along the margin of these pools where they develop rapidly and attempt to transform before the aquatic site dries.

The family is the sister taxon to Dendrobatidae (Pyron and Wiens 2011), a radiation centered in Central and South America. Because of these relationships, the family Bufonidae is thought to have originated in South America, radiated across the globe, and then returned new lineages to the New World by crossing the Bering Land Bridge (Pramuk et al. 2007). North American Toads (*Anaxyrus*) and Central American Toads (*Incilius*) are the only extant genera native to the United States, and both are derived from the re-invasion of the New World. Both are found in Alabama.

KEY TO THE GENERA OF BUFONIDAE OF ALABAMA

1a A parietal crest present on top of head; tadpole with a series of light spots along top of tail musculature.

Genus *Incilius*—Central American Toads . . . page 53

From left to right:

Dorsal view of head of Central American Toad (*Incilius*) showing interorbital, postorbital, and parietal crests.

Lateral view of Central American Toad (*Incilius*) tadpole showing light spots along tail musculature.

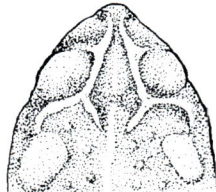
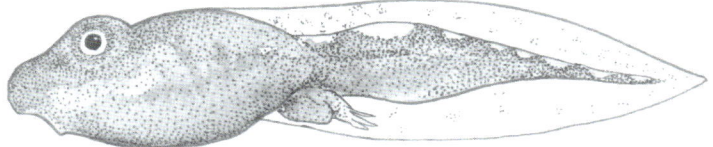

1b No parietal crest on top of head; tadpole lacking a series of light spots along top of tail musculature.

Genus *Anaxyrus*—North American Toads . . . page 57

Dorsal view of head of American Toad (*Anaxyrus americanus*) showing interorbital and postorbital crests but no parietal crests.

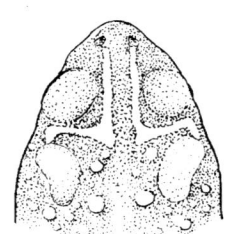

Central American Toads
Genus *Incilius* (Cope, 1863)

We follow Mendelson et al. (2011) in recognizing *Incilius* as a genus of toads of Central American origin. The genus consists of 39 species distributed from the Pacific coast of Ecuador to the southern US, of which one is found in Alabama. Although no morphological feature is diagnostic of the genus, members of it tend to have flattened heads and bodies with sharply pointed warts. *Incilius* is the sister genus to *Anaxyrus* (Pauly et al. 2009), a North American radiation of toads. Our use of this genus name follows that of others (e.g., Pramuk et al. 2007) who have accepted the notion that the breakup of the large, cosmopolitan genus *Bufo* assists in understanding patterns of evolutionary and biogeographic history. However, Pauly et al. (2009) have advocated a competing taxonomy in which the North American Toads are retained in the genus *Bufo*, the genus name used for decades for these animals, along with most other members of the family. In their classification, three subgeneric categories are recognized for the New World toads, including a Central American clade that is identical in content to *Incilius*.

Adult Gulf Coast Toad (*Incilius nebulifer*), Montgomery County, AL.

Gulf Coast Toad
Incilius nebulifer (Girard, 1854)

DESCRIPTION Gulf Coast Toads are medium-to-large-sized bufonids, attaining a maximum snout–vent length of about 5 in (125 mm). The head and body appear to be slightly flattened from top to bottom. The skin of this species has numerous spine-tipped warts, with a noticeably enlarged series of warts extending from the posterior tip of each parotoid gland, along each side of the body, to the groin. The parotoid gland is roughly triangular, being broader anteriorly and narrower posteriorly. The venter is uniform white. The dorsum is brown in ground color with a yellowish-white middorsal stripe from the back of the head to the cloacal opening and a pair of wide dorsolateral stripes from the back of the head to the upper surfaces of the thigh, shank, and foot. Edges of the light dorsal stripes are diffuse. The head is a uniform greenish tan, with the cranial crests accentuated by being black or dark brown. Tadpoles of Gulf Coast Toads are small, uniform dark gray, and have a series of white spots along the upper surface of the tail.

In Alabama, Gulf Coast Toads are easily confused with the Southern Toad (*Anaxyrus terrestris*). But the two can be distinguished by the presence of a parietal crest in Gulf Coast Toads and its absence in Southern Toads. Gulf Coast Toads are known to hybridize with Fowler's Toad (*A. fowleri*; Dundee et al. 1996), but no such hybridization is known from Alabama.

ALABAMA DISTRIBUTION This species was first detected in 2016 in a riparian forest in Montgomery County (Birkhead et al. 2017). Subsequent detections have been concentrated in relatively new housing developments in the area. The species is documented from the Lower Coastal Plain of Mississippi, with the main portion of its range encompassing Louisiana and eastern Texas. The fact that the species has not been observed elsewhere in the Lower Coastal Plain of Alabama is problematic. The species is conspicuous in its breeding habits, and so it seems unlikely that these toads avoided discovery by previous attempts to summarize Alabama's herpetofauna (Mount 1975). Similarly, it seems unlikely that the species dispersed on its own to Montgomery County. Instead, human-assisted dispersal, perhaps via transport as a stowaway in nursery plants seems the most likely explanation for the occurrence of this species in Alabama.

HABITS The Gulf Coast Toad inhabits hardwood forests, riparian areas, and urban sites during most of the year, tending to avoid pinelands. Because they survival well in urban settings, this species may be found feeding on insects attracted to lights at night. These toads migrate to breeding sites after rains from February to August (Goldberg 2017), using roadside ditches, ponds, river flood plains, and impoundments.

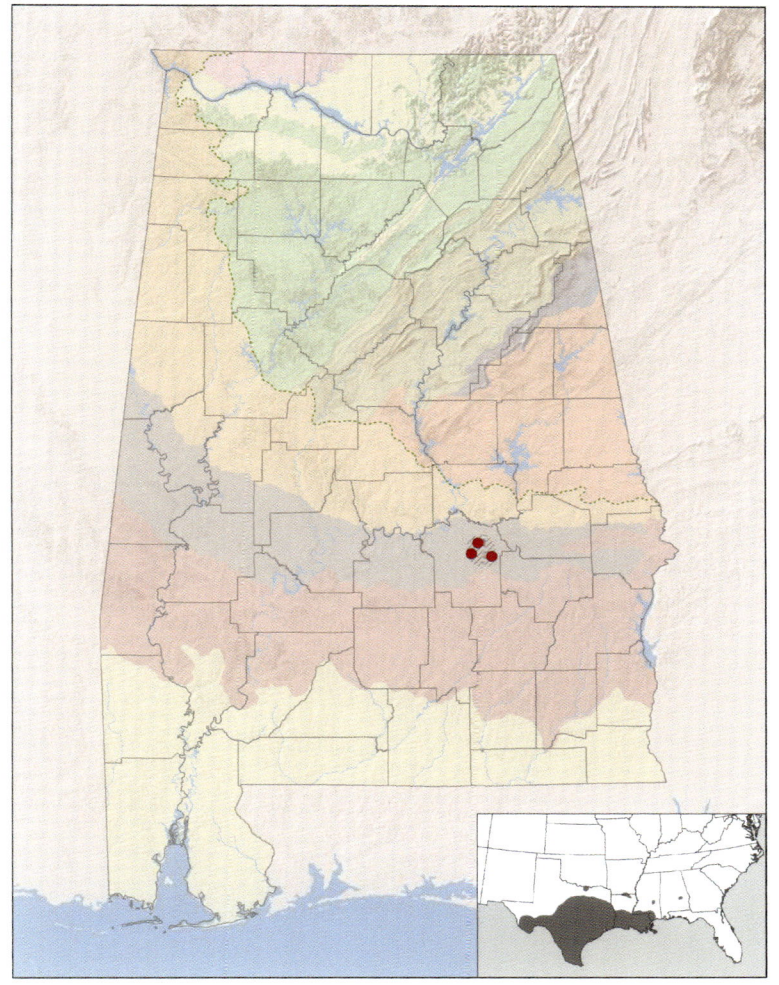

Distribution of Gulf Coast Toad (*Incilius nebulifer*). The presumed range of the species in Alabama is indicated by hatching. Solid dots indicate localities of specimens or photographs examined by the authors or ADCNR/Natural Heritage Program occurrence records believed to be valid. Inset map depicts approximate range in the United States.

Males give a raspy, low-pitched, pulsed call that lasts 4–6 seconds and is used to attract mates. The call is similar to that of Southern Toads, but lower in pitch and slower in pulse rate. Females select shallow water and exude two strings of up to 20,000 eggs that are fertilized externally by the male. Tadpoles transform in 20–30 days. Tadpoles scrape algae or carcasses of dead tadpoles, and adults eat arthropods. Individuals may move their colorful toe tips to attract prey items (Erdmann 2017). Tadpoles are unpalatable to sunfish, a feature that may help them invade new areas (Adams et al. 2011).

CONSERVATION AND MANAGEMENT Gulf Coast Toads are common in Louisiana and Texas, where they receive no special protection. Abundance of this species does not appear to be altered by forestry activities (Foley 1994), and so these toads appear to be resilient to human activities. Surveys for this species within Alabama are needed to refine our understanding of its distribution within the state. Regardless of the outcome of such surveys, Alabama populations represent the periphery of the range of the species. It is possible that the species recently invaded Alabama, and surveys should be designed to test this possibility. Where this species shares breeding sites with Fowler's Toad, Gulf Coast Toad tadpoles develop normally while those of Fowler's Toad transform at a smaller size (Vogel and Pechmann 2010). Thus, Gulf Coast Toads may outcompete Fowler's Toad. Additionally, invasive plants, like Chinese Tallow Tree (*Triadica sebifera*), do not affect growth of Gulf Coast Toads in the same way that these plants affect other native anurans (Cotton et al. 2012). Chytrid fungal disease (*Batrachochytrium dendrobatidis*) can be induced in a lab setting but is not known for free-ranging individuals (Brannelly 2014).

TAXONOMY We follow Mendelson et al. (2015) in considering this species to be distinct from the Southern Gulf Coast Toad (*Incilius valliceps*).

North American Toads

Genus *Anaxyrus* (Tschudi, 1845)

Anaxyrus is a recently elevated genus that contains most of the bufonids of North America (Frost et al. 2006). The genus consists of 23 species, of which 4 are found in Alabama. It is the sister genus to *Incilius*, a Central American radiation of toads that includes some forms from the southern United States (Pauly et al. 2009). Our use of this genus name follows that of others (e.g., Pramuk et al. 2007) who have accepted the notion that the breakup of the large, cosmopolitan genus *Bufo* assists in understanding patterns of evolutionary and biogeographic history. However, Pauly et al. (2009) have advocated a competing taxonomy in which the North American Toads are retained in the genus *Bufo*, the genus name used for decades for these animals, along with most other members of the family. In their classification, three subgeneric categories are recognized for the New World toads, including a North American clade that is identical in content to *Anaxyrus*.

KEY TO THE SPECIES OF *ANAXYRUS* OF ALABAMA

1a Adult snout–vent length less than 1.4 in (35 mm); dorsum with paired dark spots separated by a light median stripe; first upper labial tooth row of tadpole with a median gap; tail musculature of tadpole with diffuse black bars separating regions of light brown.

   ***Anaxyrus quercicus*—Oak Toad . . . page 60.**

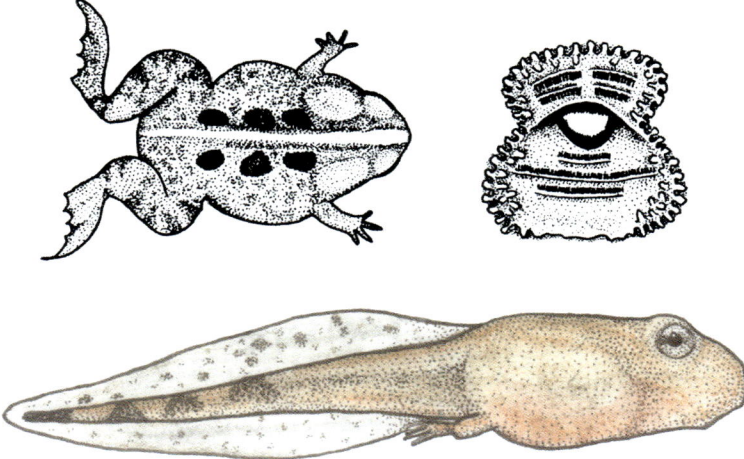

Clockwise from upper left:

Dorsal view of Oak Toad (*Anaxyrus quercicus*).

Ventral view of Oak Toad (*Anaxyrus quercicus*) tadpole mouth parts showing median gap in first upper labial tooth row.

Lateral view of Oak Toad (*Anaxyrus quercicus*) tadpole.

1b Adult snout–vent length greater than 1.6 in (40 mm); dorsum variously marked with spots but lacking paired spots separated by a median light stripe; first upper labial tooth row of tadpole uninterrupted; tail musculature lacking a series of light and dark bands; **go to 2.**

Ventral view of Fowler's Toad (*Anaxyrus fowleri*) tadpole mouth parts showing uninterrupted first upper labial tooth row.

2a Postorbital crest touching parotoid; large dorsal spots usually with three or more warts; tadpole mottled, with rounded snout and lower half of tail fin about 50 percent as wide as tail musculature.

Anaxyrus fowleri—Fowler's Toad . . . page 64.

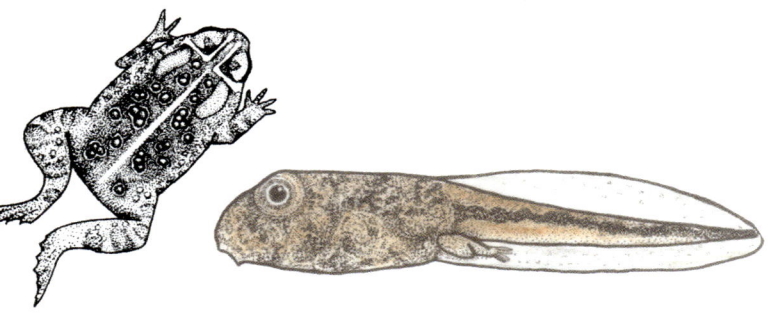

From left to right:

Dorsal view of Fowler's Toad (*Anaxyrus fowleri*) showing postorbital crest touching anterior edge of parotoid gland and dark dorsal blotches with three or more warts.

Lateral view of Fowler's Toad (*Anaxyrus fowleri*) tadpole with lower caudal fin about 50 percent as wide as caudal musculature.

2b Postorbital crest not touching parotoid but having a backward-projecting spur that does; dorsal spots, if present, usually with 1–2 warts (see illustrations with couplet 4); tadpole black with sloped snout and lower half of tail fin about 25 percent as wide as tail musculature; **go to 3.**

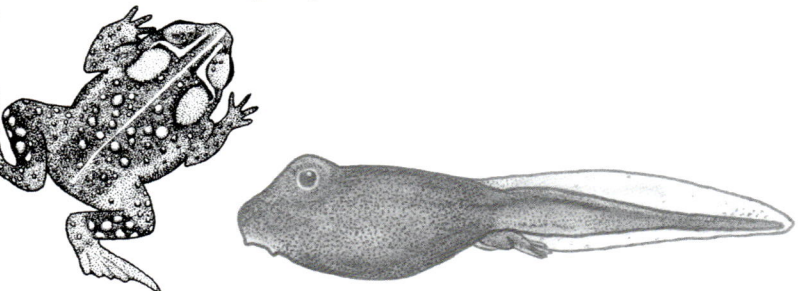

From left to right:

Dorsal view of American Toad (*Anaxyrus americanus*) with backward spur connecting postorbital crest to anterior edge of parotoid gland.

Lateral view of American Toad (*Anaxyrus americanus*) tadpole showing width of ventral portion of caudal fin about 25 percent as wide as tail musculature.

3a Supraorbital crest ending posteriorly in pronounced knoblike protuberance; venter immaculate white; tadpole with light (golden) oblique line from behind each eye to side of body and distance from eye to tip of snout about three times eye width.

Anaxyrus terrestris—**Southern Toad . . . page 68.**

From left to right:

Dorsal view of Southern Toad (*Anaxyrus terrestris*).

Lateral view of Southern Toad (*Anaxyrus terrestris*) tadpole.

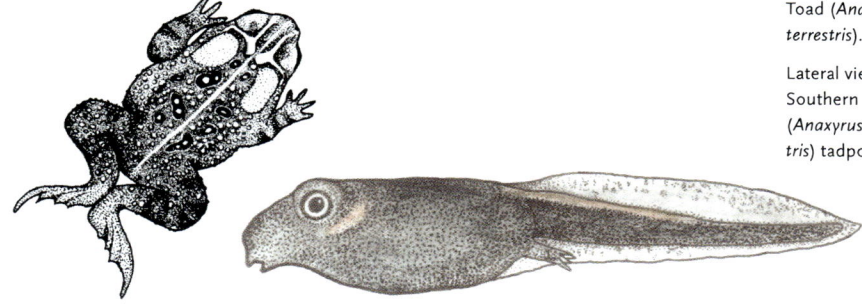

3b Supraorbital crest not ending in knoblike protuberance; venter often with dark spots; tadpole with numerous small golden dots concentrated along top of tail musculature and distance from eye to tip of snout about 1.5 times eye width.

Anaxyrus americanus americanus—**Eastern American Toad . . . page 73.**

From left to right:

Dorsal view of Eastern American Toad (*Anaxyrus americanus americanus*).

Lateral view of Eastern American Toad (*Anaxyrus americanus americanus*) tadpole.

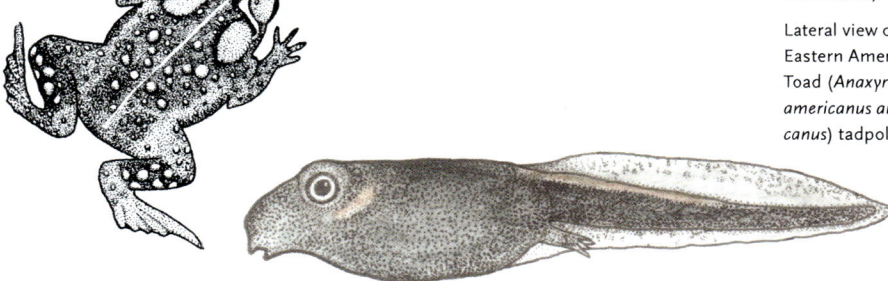

Adult Oak Toad
(*Anaxyrus quer-*
cicus), Wakulla
County, FL.

Oak Toad

Anaxyrus quercicus (Holbrook, 1840)

DESCRIPTION The smallest North American member of its family, Oak Toads attain a maximum snout–vent length of about 1.4 in (35 mm). Cranial crests are low, forming a broad, flat, raised area across the front of the snout (boss). The parotoid glands are large and somewhat divergent posteriorly instead of being parallel to the body. The dorsal ground color is gray or brown to almost black with a narrow white or yellow median stripe. There are paired dark dorsal blotches, each bisected by the median stripe, and the dorsum typically is rather evenly warty. Tadpoles of Oak Toads are small, golden brown in color, and have tail musculature that is golden brown with a series of diffuse dark bars toward the tip separating light and dark patches.

The small size and lack of postorbital crests of Oak Toads will distinguish this species from adults of the other three species of *Anaxyrus* in Alabama. Because the cranial crests of juveniles of Alabama's other three species are weakly developed, they may be confused with Oak Toads. However, these three species will have smaller parallel-oriented parotoid glands, unpaired dark dorsal spots, and no elevated boss across the forehead. Tadpoles of Alabama's other *Anaxyrus* species also differ from those of Oak Toads in being darker in ground color and lacking diffuse dark bars on the tail musculature.

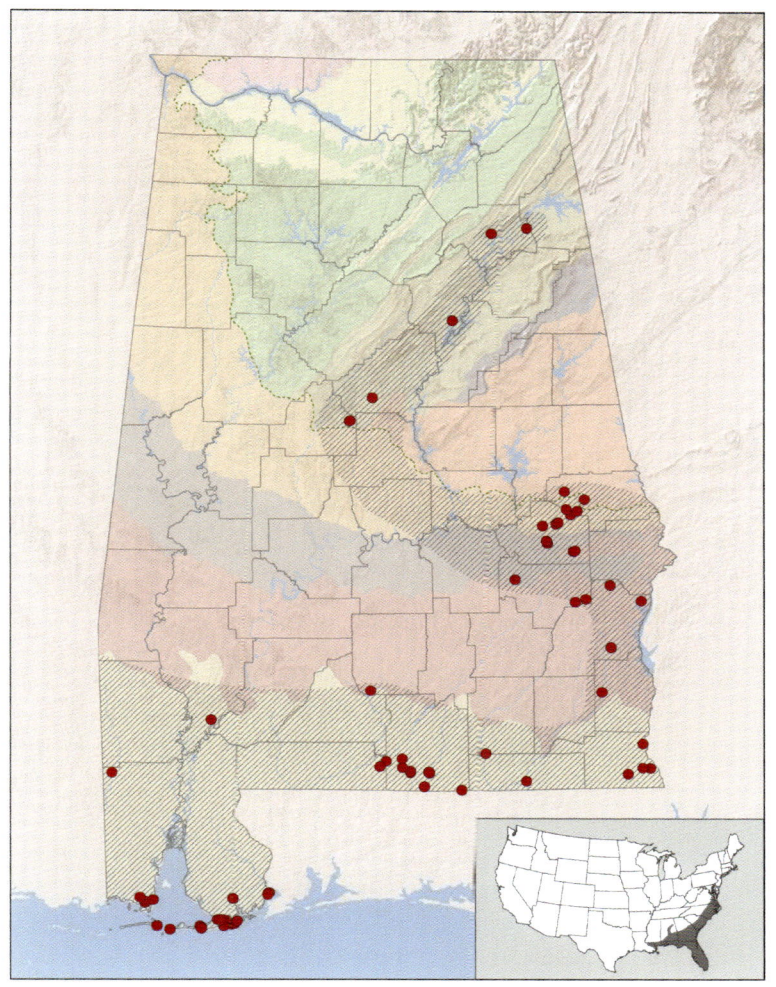

Distribution of Oak Toad (*Anaxyrus quercicus*). The presumed range of the species in Alabama is indicated by hatching. Solid dots indicate localities of specimens or photographs examined by the authors or ADCNR/Natural Heritage Program occurrence records believed to be valid. Inset map depicts approximate range in the United States.

ALABAMA DISTRIBUTION Oak Toads have a patchy distribution across the Lower Coastal Plain and are found elsewhere in the Coastal Plain in areas where the soil is relatively sandy. The range extends across the Fall Line into the Ridge and Valley region, northeastward at least to eastern Etowah County.

HABITS This miniature toad tends to be fossorial, and in Alabama it is seldom encountered except during the breeding season. However, like other members of the genus, it is presumed to spend most of its active time in upland habitats where feeding and growth occurs. Surface activity of adults occurs May through September and peaks during heavy rainfall from June through August (Greenberg and Tanner 2005b), when adults migrate to temporary wetlands to breed. The call, a high-pitched "peep, peep . . . peep," with each note slurred downward slightly at the end, closely resembles the peeping of a baby chicken. It is usually issued from a clump of grass or other sheltered situation around the edge of a flooded roadside ditch, borrow pit, or other small pond or pool. During such periods, the call is often heard by day. Females appear to be attracted to larger males in breeding choruses (Wilbur et al. 1978), allow amplexus to occur, and then lay eggs in small strands or bars, with 3–6 eggs per bar. These are fertilized externally by the male. Up to 500 eggs are laid per female (Volpe and Dobie 1959). Tadpoles form aggregations along the margins of these pools. As with all toads, Oak Toads produce noxious skin toxins that are transferred by the female to the eggs, making the eggs toxic (Brodie et al. 1978).

The numbers of adults that appear at breeding sites differ each year and likely decrease during droughts (Dodd 1994). Complete reproductive failure is common at most breeding sites (Greenburg and Tanner 2005b). Despite such failures, Oak Toads continue to return to each breeding site (Dodd 1994). Juvenile recruitment is greater in ponds that hold water for longer time periods (Greenberg and Tanner 2005b), and so certain wetlands may maintain populations more readily than others. Only about 2 percent of toads move to a different pond to breed between years, and when they move to a new pond, they only move about 110 yd (100 m) (Greenberg and Tanner 2005b). Habitat quality around breeding sites is important in maintaining populations, with upland, fire-maintained Longleaf Pine (*Pinus palustris*) forests being crucial (Delis et al. 1996).

Oak Toads consume small arthropods, especially ants and beetles. Tadpoles consume periphyton and phytoplankton but also are known to consume carcasses of other tadpoles.

CONSERVATION AND MANAGEMENT Oak Toads are designated by ADCNR as a Priority 3 (species of moderate conservation concern; Shelton-Nix 2017), because the species has such a patchy distribution and because populations appear to be vulnerable to extirpation. Proper management of Longleaf Pine forests on deep sandy soils will be key to retaining this species in the state. Frequent low-intensity fire, especially during the growing season, will generate the grass-dominated understory preferred by this species, as will thinning of dense pine stands. Protection of natural depressional wetlands and creation of artificial ones may be needed, especially on public lands, such as Fort Rucker, Geneva State Forest, Conecuh National Forest, and Perdido River Longleaf Hills Tract. Oak Toad conservation will require preserving aggregations of breeding sites to make sure that at least one pond produces recruits during each year and to allow for inter-pond movements (Greenberg and Tanner 2005b). This species is retained at agricultural lands used as pastures (Babbitt and Tanner 2000), but upland pine forests must be near such sites (Babbitt et al. 2006). Thus, the species is an indicator of high-quality pine wetlands and is quickly extirpated from urban landscapes (Guzy et al. 2012).

TAXONOMY Phylogenetically, this species diverges near the base of the tree for *Anaxyrus* and represents an isolated branch of eastern origin nested within other basal species that are largely distributed across the western United States (Pauly et al. 2004). No subspecific variation is recognized within this taxon.

Adult Fowler's Toad
(*Anaxyrus fowleri*),
Bibb County, AL.

Fowler's Toad
Anaxyrus fowleri (Hinckley, 1882)

DESCRIPTION Fowler's Toads are medium-sized bufonids, attaining a maximum snout–vent length of about 3.3 in (85 mm). These animals have warty skin, and the warts on the tibia are of similar size to those on the thigh. Parotoid glands are present and obvious. The pupil of the eye is horizontally elliptical. In dorsal coloration, these toads are extremely variable, ranging from light gray to brick red, typically with large dorsal dark spots that have three or more warts in each spot. The venter is light except for an occasional dark breast spot. Cranial crests are not prominent in this species, and the supraorbital crests do not end in knoblike protuberances where they join the postorbital crests. Instead, each postorbital crest contacts a parotoid gland directly. Tadpoles of Fowler's Toads are small with a rounded snout, have a body that is mottled with dark brown, and possess a dark brown stripe along the tail musculature with clear, unspotted tail fins.

In Alabama, Fowler's Toads are easily confused with Eastern American Toads (*A. americanus americanus*) and Southern Toads (*A. terrestris*). Hybrids of Eastern American Toads and Fowler's Toads are found in the northern third of the state (Masta et al. 2002) and Fowler's Toad and Southern Toads hybridize freely in the southern portion of the state (Mount 1975). Throughout the state, individuals with three or more warts per dark blotch, no enlarged tibial warts, no enlarged protuberance at the intersection of the supraorbital and postorbital crests, and direct contact between the postorbital crest to the parotoid gland

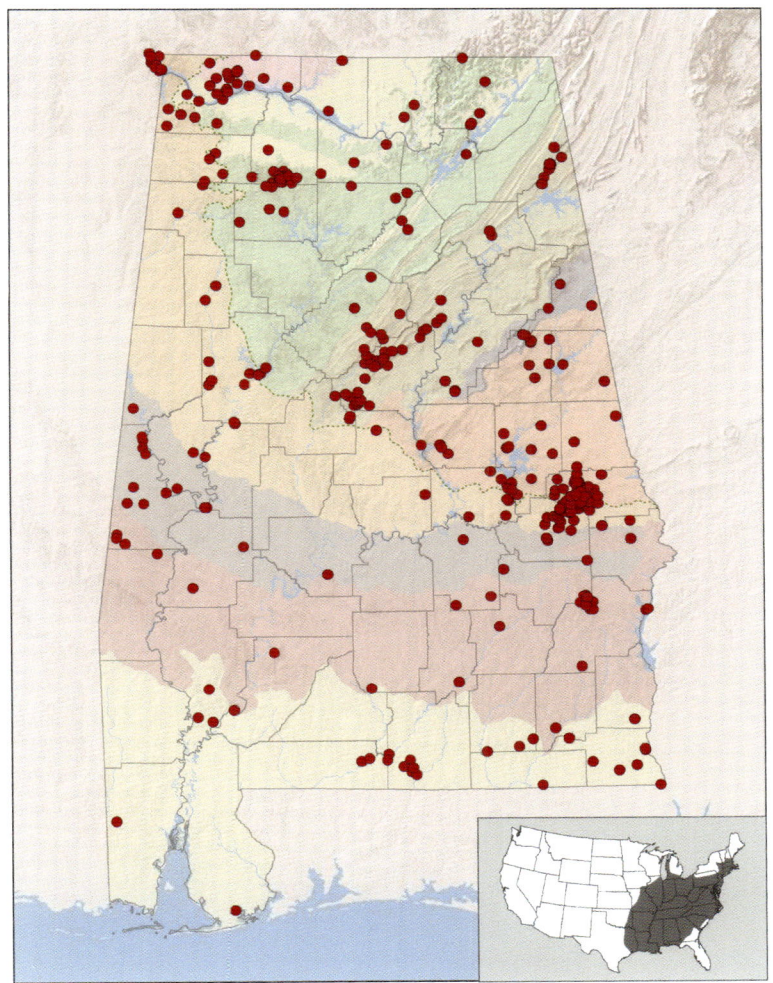

Distribution of Fowler's Toad (*Anaxyrus fowleri*). Solid dots indicate localities of specimens or photographs examined by the authors or ADCNR/Natural Heritage Program occurrence records believed to be valid. Inset map depicts approximate range in the United States.

are most likely to be Fowler's Toad. But, patterns of hybridization assure that some of these individuals will belong to a different species. Fowler's Toad tadpoles are most like tadpoles of the Southern and Eastern American Toads, which are black rather than brown.

ALABAMA DISTRIBUTION Fowler's Toad is found throughout the state. But this species is not common in the Lower Coastal Plain, being largely restricted to riparian zones of major rivers in this region.

HABITS Fowler's Toad occurs in most, if not all, of our terrestrial habitat types. It thrives in residential areas of most cities and towns, as well as in rural districts. During winter months (November–March) it burrows underground at upland sites away from breeding wetlands. Fowler's Toad emerges later in the season than does the Eastern American Toad and slightly earlier than Southern Toads. Breeding activity begins in March or April, peaks in May, and may continue into August. Fowler's Toad breeds during both dry and wet weather. The breeding sites selected by Fowler's Toad are typically of a more permanent nature than those used by Southern Toads and Eastern American Toads. Breeding sites for Fowler's Toads include lakes, farm ponds, rivers, creeks, and drainage ditches. The call note is a short, nasal "waah," issued from a spot near the water's edge. Males typically outnumber females during bouts of breeding despite an overall 1:1 sex ratio for the entire population (Green 2013). The eggs, up to 10,000 in number, are laid in strings submerged along the bottom. Tadpoles of this species are found in aggregations along the margins of these pools. Adults of this species produce noxious skin secretions, and these are transferred by the female to the eggs, making the eggs toxic (Brodie et al. 1978).

Once breeding activities have ended, adults and transformed toadlets migrate to upland sites where adults tend to burrow in loose soil or leaf litter during the day and emerge to feed at night. However, juveniles may be found active both day and night. Fowler's Toad feeds almost exclusively on insects and other arthropods, and because the species survives well in urban settings, adults may be found feeding on insects attracted to lights at night. Tadpoles consume periphyton and phytoplankton but also are known to consume carcasses of other tadpoles.

CONSERVATION AND MANAGEMENT Fowler's Toads are abundant and widespread within the state. Therefore, the species receives no regulatory protection in Alabama. Reproductive sites include farm, park, and golf course ponds created by humans, demonstrating that these toads coexist with humans easily. In fact, these toads are among the earliest to invade new human-constructed wetlands (Birx-Raybuck et al. 2009). For these reasons no specific management activities are required to retain them. This species, over the long term, shows no tendency to breed earlier or to have breeding associated with warm temperatures, features expected if global climate change is altering anuran reproduction (Blaustein et al. 2001). But long-term monitoring from the northeastern US suggests decreasing occupancy for this species over time (Weir et al. 2014). Additionally, site occupancy is negatively affected by the presence of deciduous trees, hayfields, human developments, and pesticide use (Jones and Tupper 2015). Population declines have been associated with expansion of invasive plant species (Greenberg and Green 2013). Where Fowler's Toads share breeding sites with Gulf Coast Toads (*Incilius nebulifer*), tadpoles of Fowler's Toads transform at smaller body sizes than normal (Vogel and Pechmann 2010), suggesting that Fowler's Toads are outcompeted by Gulf Coast Toads. Chytrid fungal disease (*Batrachochytrium dendrobatidis*) has been detected within populations of Fowler's Toads (Davis et al. 2012).

TAXONOMY Fontenot et al. (2011) documented a monophyletic nuclear lineage with a whining male advertisement call that contained Fowler's Toad, East Texas Toad (*A. velatus*), and Woodhouse's Toad (*A. woodhousii*). We consider Alabama specimens clustered along this lineage to be Fowler's Toads. Five such Alabama specimens document that this species is present in the state. This species has no subspecies, but it hybridizes with Eastern American Toads, Southern Toads, and Gulf Coast Toads, creating uncertainty when external morphology alone is used to diagnose this complex of species. A South Carolina specimen with Fowler's Toad external features possesses a nuclear genome that unites it with Southern Toads; a Mississippi specimen with American Toad (*A. americanus*) external features possesses the nuclear genome of Fowler's Toad; Alabama and Florida specimens with Southern Toad external morphology possess nuclear genomes assigning the specimens to Fowler's Toad. Nevertheless, based on its nuclear genome, Fowler's Toad is a distinct historical lineage that is the sister lineage to American Toads (Fontenot et al. 2011).

Southern Toad

Anaxyrus terrestris (Bonnaterre, 1789)

DESCRIPTION Southern Toads are medium-sized bufonids, attaining a maximum snout–vent length of about 3.1 in (80 mm). The skin of this species is warty, and the warts often are spine tipped. Parotoid glands are present and are obvious. The pupil of the eye is horizontally elliptical. In dorsal coloration, this species is extremely variable, ranging from various shades of gray to brick red and possessing dark gray spotting or mottling. The venter is uniform white. Cranial crests are prominent with the interorbital crests projecting beyond their junction with the postorbital ridge, creating prominent knoblike protuberances. The postorbital ridges do not contact the parotoid glands directly, but instead connect to them by a backward-projecting spur. Tadpoles of Southern Toads are small and black (except for a golden mark behind each eye) with an elongate snout. These tadpoles possess a tail with a golden stripe along the dorsal part of the muscles and fins that are suffused with dark mottling.

Adult Southern Toads are similar in appearance to Eastern American Toads (*Anaxyrus americanus americanus*), but that species possesses tibial warts that are noticeably larger than the remaining body warts and lack the enlarged protuberances at the confluence of the supraorbital and postorbital crests of Southern Toads. Eastern American

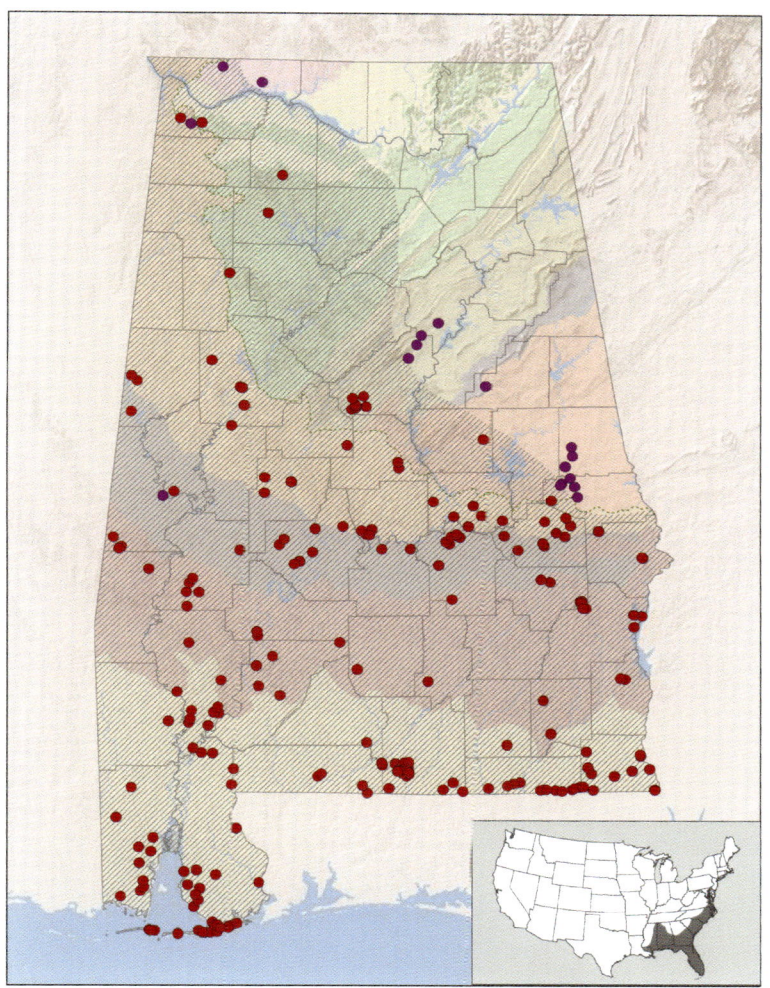

Distribution of Southern Toad (*Anaxyrus terrestris*). The presumed range of the species in Alabama is indicated by hatching. Solid red (morphologically pure) and purple (morphological hybrids with Eastern American Toads [*A. americanus americanus*]) dots indicate localities of specimens or photographs examined by the authors or ADCNR/Natural Heritage Program occurrence records believed to be valid. Inset map depicts approximate range in the United States.

Toad tadpoles also are like tadpoles of the Southern Toad, but Eastern American Toad tadpoles lack a golden spot behind the eye and a stripe on the dorsum of the tail musculature.

ALABAMA DISTRIBUTION This species is common throughout the Coastal Plain of the state. Based on Chivers (2016), morphologically pure Southern Toads are also found throughout the Ridge and Valley, and Appalachian Plateaus regions. These specimens likely are hybrids because only the calls of Eastern American Toads are heard in these regions. Specimens from Tuscaloosa County northward within the

Fall Line Hills are most similar, morphologically, to Southern Toads but have reduced cranial protuberances. We infer that the western portion of the state from Sumter to Lauderdale Counties is a wide zone of intergradation between Southern and Eastern American Toads.

HABITS This is the most common toad in the Lower Coastal Plain and is found in approximately equal numbers with Fowler's Toad in other Coastal Plain provinces. It seems especially abundant in areas with friable soil. The fact that the range of Southern Toads crosses the Fall Line from the Coastal Plain into the Ridge and Valley region and Appalachian Plateaus, but stops within the Piedmont, may be related to substrate requirements. Adults and juveniles spend much of the active season in upland pine habitats where they feed and grow. This species burrows in loose soil and leaf litter during the day, emerging at night. During winter months (November–March) the species remains dormant in burrows, emerging in March to breed. Breeding in this species usually begins somewhat later than in Eastern American Toads or Fowler's Toad (*A. fowleri*) and is confined mostly to wet periods from mid-March to early June. Breeding may occur in small permanent ponds, woodland pools, or flooded depressions to which adults migrate. This species does not breed in creeks or rivers, as Fowler's Toad often does. Adults migrate at night during and immediately after rain showers (Todd and Winne 2006). The call is a high-pitched, drawn-out trilled whistle, about half an octave higher than that of Eastern American Toads. Large-male advantage in breeding aggregations results from the fact that large males can dislodge small males that amplex females, but small males do not attempt to dislodge large males (Lamb 1984). Once a female is amplexed by a male, the female deposits eggs under water as two long strings, one from each ovary, and the male fertilizes them externally. After eggs are deposited, females migrate back to upland sites for the remainder of the active season. Males remain within or near the breeding site until breeding activity ceases and then migrate to upland areas. Tadpoles are found in aggregations along the margins of temporary pools. All individuals produce noxious skin secretions, and these are transferred by the female to the eggs, making them toxic (Brodie et al. 1978). Tadpoles tend to transform in large groups, remaining near the breeding site for several weeks before eventually migrating to upland areas.

Southern Toads feed almost exclusively on insects and other arthropods, especially ants, beetles, and spiders (Punzo 1992). Tadpoles consume periphyton and phytoplankton but also are known to consume carcasses of other tadpoles.

In Alabama, the advertisement calls of male Fowler's Toads and Southern Toads become more divergent from each other where the two species are found in sympatry than where they occur in allopatry (Leary 2001), suggesting selective pressure to reduce hybridization between the two species. Additionally, the release call of male Fowler's Toads and Southern Toads become more alike when the two species are in sympatry than when they are in allopatry (Leary 2001).

CONSERVATION AND MANAGEMENT Southern Toads are abundant and widespread in the southern third of the state. Therefore, the species receives no regulatory protection in Alabama. The reproductive sites include temporary wetlands created by human activities, and these toads frequently live in garden areas, with adults visiting outdoor house lights to search for food. For these reasons, populations seem secure and occupancy is not affected by urbanization (Guzy et al. 2012). Additionally, occupancy increases around wetlands in areas with more impervious surfaces (Alix et al. 2014a). However, frequent, low-intensity prescribed fire produces habitat characteristics preferred by Southern Toads (open-canopied pine forest with an understory dominated by herbaceous plants). This species typically emerges in great numbers at night immediately after such fires.

Environmental changes that are detrimental to Southern Toads are known. Coal ash, a product of coal-burning power plants that is rich in heavy metals, is frequently disposed of by spreading it in remote areas. This may create population sinks for Southern Toads because complete mortality of tadpoles may occur at wetland sites containing such ash. This results because of direct mortality caused by the ash and indirect effects through reduction of algae used as food by these tadpoles (Rowe et al. 2001). Similarly, agro-forestry activities are known to have deleterious effects on Southern Toads. Clear-cut areas are occupied by these toads but they avoid areas of thick regrowth during recovery from clear-cutting (Todd et al. 2009). Juveniles in clear-cut areas experience increased mortality and slower growth, factors that could make local populations inviable if the clear-cut area is large enough

(Todd and Rothermel 2006). Nitrate runoff from application of fertilizers alters tadpole growth in Southern Toads, indicating that this is a significant environmental stressor (Edwards et al. 2006). Deleterious effects of salinity and application of the insecticide carbaryl on tadpole growth and survival are exacerbated at wetland sites where both factors are increased (Wood and Welch 2015). Uptake of environmental copper by tadpoles reduces their survival (Lance et al. 2013) and maternal transfer of metal pollutants to eggs reduces reproductive success (Metts et al. 2013). Finally, chytrid fungal disease has been detected in Southern Toads but has not been associated with mass mortality (Rizkalla 2010).

Adult Eastern American Toad (*Anaxyrus americanus americanus*), Randolph County, AL.

Eastern American Toad

Anaxyrus americanus americanus (Holbrook, 1836)

DESCRIPTION Eastern American Toads are medium-to-large-sized bufonids, attaining a maximum snout–vent length of about 4.3 in (110 mm). The skin of this species is warty with the warts usually lacking spine tips, except in large females. The warts on the dorsal surface of the tibia are larger than those of the rest of the leg. Parotoid glands are present and obvious, and the pupils of the eyes are horizontally elliptical. The dorsal ground color is variable, ranging from gray to brown or reddish, and may be uniform or variously spotted and mottled with dark gray blotches. Most dark dorsal blotches encompass only one or two warts. The venter is light, occasionally with a dark spot at the center of the breast. The cranial crests are prominent on adults with the supraorbital crests ending at their junction with the postorbital ridges. The postorbital ridges do not contact the parotoids directly, but instead do so by backward-projecting spurs. Tadpoles of Eastern American Toads are small and uniform black, have a rounded snout, and possess a tail with clear, unspotted fins.

In Alabama, Eastern American Toads are easily confused with both Fowler's Toad (*A. fowleri*) and Southern Toads (*A. terrestris*). Hybrids of Eastern American Toads and Fowler's Toad are found in the northern third of the state (Masta et al. 2002), and hybrids between Eastern

American Toads and Southern Toads are found in central portions of the state (Mount 1975). In northern Alabama, individuals with one or two warts per dark blotch, enlarged tibial warts, no enlarged protuberance at the intersection of the supraorbital and postorbital crests, and a spur-like crest connecting the postorbital crest to the parotoid gland are likely to be Eastern American Toads. But specimens with some features of Fowler's Toad (three warts per dark blotch, tibial warts of similar size to those of rest of leg, and postorbital crest contacts parotoid gland) also may be Eastern American Toads. Specimens from central Alabama that possess a slightly enlarged knoblike protuberance at the junction of the supraorbital and postorbital crests are likely to be hybrids between Eastern American Toads and Southern Toads. Eastern American Toad tadpoles are most like tadpoles of the Southern Toad, which are black with an elongate snout, golden spot behind each eye, a golden stripe on the dorsum of the tail musculature, and diffuse markings on the tail fin.

ALABAMA DISTRIBUTION This subspecies is common throughout the Piedmont and Ridge and Valley regions, is missing from Lookout and Sand Mountains of the Appalachian Plateaus, and is present in the complex mountains north of the Tennessee River in Jackson and Madison Counties. Based on Chivers (2016), morphologically pure Eastern American Toads are found as far west as the Fall Line Hills of Bibb County and as far south as Montgomery County. However, extensive regions of hybridization with Southern Toads occur in western regions of the Piedmont and Ridge and Valley formations (Mount 1975) and likely along the Fall Line Hills.

HABITS The Eastern American Toad is the first *Anaxyrus* to emerge from hibernation in Alabama, and it may be seen on the roads on warm, rainy nights as early as mid-January, along with Southern Leopard Frog (*Lithobates sphenocephalus*), Spring Peeper (*Pseudacris crucifer*), Upland Chorus Frog (*P. feriarum*), and Mountain Chorus Frog (*P. brachyphona*). Adults overwinter in burrows in loose soil in forested upland areas. Breeding activity usually gets underway in late January or February, when adults migrate to breeding wetlands, with the peak of the season occurring in mid-March or, in extreme northern Alabama, mid-April. In Alabama, this species breeds as late as early May. Breeding sites include farm ponds, floodplain pools, and flooded roadside ditches. The male calls from banks near the edge of water,

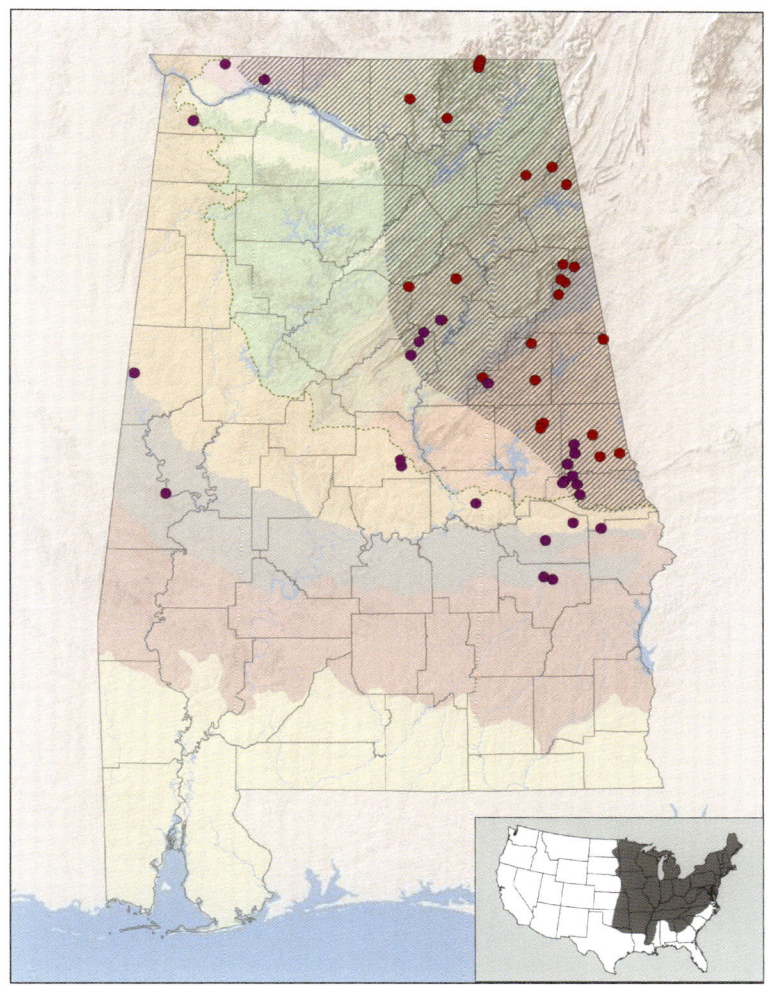

Distribution of Eastern American Toad (*Anaxyrus americanus americanus*). The presumed range of the subspecies in Alabama is indicated by hatching. Solid red (morphologically pure) and purple (morphological hybrids with Southern Toad [*A. terrestris*]) dots indicate localities of specimens or photographs examined by the authors or ADCNR/Natural Heritage Program occurrence records believed to be valid. Inset map depicts approximate range in the United States, with dark shading indicating the range of Eastern American Toad and light shading indicating the range of all other subspecies.

emitting a resonant, drawn-out trilled whistle, about half an octave lower than that of Southern Toads and having a slower pulse rate. Females prefer males with lower dominant frequencies (low call pitch). Males calling alone call at a higher pitch than when that same male calls with a nearby neighbor, and the negative relationship between male body size and call dominant frequency observed in most frogs becomes more pronounced in choruses of Eastern American Toads than for isolated calling males (Howard and Young 1998). These data suggest that calls of deeper pitch are more energetically expensive to produce and that competition for mates is more intense in calling choruses of Eastern American Toads. Males remain in a calling chorus for about one week (Sullivan 1992), breeding as often as possible before migrating to upland wooded areas to spend the rest of the active season feeding and growing. Eggs are laid in strings on the bottom of temporary pools. After females have deposited eggs, they migrate to upland woods to feed for the rest of the active season. These females may move from 437 yd (400 m) to 0.6 mi (1 km) from breeding to overwintering sites (Forester et al. 2006).

Tadpoles are found in aggregations along the margins of these pools. All individuals produce noxious skin secretions, and these are transferred by the female to the eggs, making these toxic (Brodie et al. 1978); they are especially unpalatable to fish predators (Kats et al. 1988). Newly hatched and premetamorphic tadpoles are unpalatable to vertebrate and invertebrate predators, but intermediate stages are palatable and rely on avoidance behavior to reduce predation (Brodie and Formanowicz 1987). Tadpoles at low density have high survival and metamorphose at a small size; as density increases survival decreases so that only a few individuals metamorphose, and these are of large size at intermediate densities but become progressively smaller as density increases (Wilbur 1977). Tadpoles and recently metamorphosed individuals form interacting aggregations, and these interactions increase size at metamorphosis (Breden and Kelly 1982), which likely improves survival of juveniles to adulthood. Juvenile toads aggregate to reduce desiccation and not to avoid predation (Heinen 1993). These small individuals remain aggregated around breeding sites for several weeks before eventually migrating to upland sites.

Eastern American Toads feed almost exclusively on insects and other arthropods. Tadpoles consume periphyton and phytoplankton but also are known to consume carcasses of other tadpoles.

Eastern American Toads appear to be greatly outnumbered within most of its Alabama range by Fowler's Toads. Eastern American Toads are limited chiefly to rural areas with some forest land while Fowler's Toads are practically ubiquitous. After mid-June, Eastern American Toads are encountered only occasionally because they are assumed to estivate during much of the summer. In contrast, both Fowler's Toad and Southern Toads are active, at least at night, throughout the summer.

CONSERVATION AND MANAGEMENT Eastern American Toads are still common and widespread within the northern half of the state. Therefore, the species receives no regulatory protection in Alabama. The reproductive sites used by this species include wetlands created by human activities, and these toads frequently live in garden areas, with adults feeding under outdoor house lights. For these reasons no specific management activities to retain them are necessary. However, in managed forests, retention of brush piles in areas that are logged is likely to reduce mortality of this species by reducing desiccation (Rittenhouse et al. 2008). Restored wetlands are readily colonized by this species (Lehtinen and Galatowitsch 2001), but calling is sensitive to road noise associated with urbanization (Cosentino et al. 2014). The insecticide carbaryl increases mortality of tadpoles of American Toads (*A. americanus*), and the herbicide atrazine increases time to metamorphosis and reduces size at metamorphosis because it reduces algae, a primary diet item of these tadpoles (Boone and James 2003). Chytrid fungal disease (*Batrachochytrium dendrobatidis*) has been detected in this subspecies but has not been associated with mass mortality events (Pullen et al. 2010). Mass mortality events have been associated with a high prevalence of ranavirus (Miller et al. 2011).

TAXONOMY Mount (1975) concluded that Southern Toads were sufficiently distinct in morphology and call characteristics to elevate it from subspecies status within American Toads to full species status. This left American Toads with three subspecies: Eastern American Toads (*A. americanus americanus*), Dwarf American Toads (*A. americanus charlesmithi*), and Canadian Toads [*A. americanus hemiophrys*; now considered a distinct species), with Eastern American Toads being present in Alabama. However, Mount (1975) noted that Eastern American Toads and Southern Toads in Alabama maintained their individuality only in the eastern portion of the state (Lee County), showing increasing evidence of hybridization the farther west and north one went in the state. Specimens referable to Southern Toads were recorded from Lawrence, Fayette, Lamar, and Tuscaloosa Counties of northwest Alabama, with specimens referable to Eastern American Toads being present no farther west than Madison and Shelby Counties. Mount (1975) found no evidence of Dwarf American Toads in Alabama, despite its presence in northeastern Mississippi.

Chivers (2016) used morphometric measurements of museum specimens to develop statistical models to identify Eastern American Toad and Southern Toad specimens from Alabama as well as their hybrids. That analysis generally confirmed the conclusions of Mount (1975) by documenting morphologically pure Eastern American Toads as far south as Montgomery County and as far west as Bibb County, and morphologically pure Southern Toads as far northwest as Colbert County and as far northeast as Cleburne County. Morphological hybrids were identified by Chivers (2016) from a wider range of counties than implied by Mount (1975).

Examination of the mitochondrial and nuclear genome across the entire geographic ranges of American Toads and Southern Toads failed to confirm monophyly of either species because of widespread gene flow between these taxa (Fontenot et al. 2011). Thus, there is evidence that these taxa are operating as a single individual rather than two individual species. For now, we retain Southern Toads and Eastern American Toads as separate species but encourage collection of further evidence that would challenge this hypothesis. We suspect that Southern Toads originated in peninsular Florida, where many other taxa evolved due to Pleistocene isolation as an island. Subsequent reconnection has allowed recontact with Eastern American Toads along the Fall Line in Alabama where gene flow appears to be rampant. Although known from eastern Mississippi, Dwarf American Toads are not known from Alabama, and its influence is not found within the state. However, morphological intermediates between Eastern American Toads and Southern Toads are present in the northwestern portion of the state. Additional studies are needed, especially along the Mississippi border, to better characterize the genetic makeup, morphology, and advertisement calls of toads from northwestern Alabama.

Microhylid Frogs

Family Microhylidae

This family contains frogs of a variety of sizes and lifestyles in the adult form, but with a tadpole that lacks a keratinized beak, has a unique position of the spiracle, and feeds on live prey captured at the aquatic surface by a unique projection of the lower jaw in those forms that have a free-living tadpole stage. At least 645 species are placed in this family, belonging to 51 genera. The family is the sister group to a cluster of African frog families (Arthroleptidae, Brevicipitidae, Hemisotidae, and Hyperoliidae; Frost et al. 2006; Pyron and Weins 2011), suggesting an origin on the continent of Africa. Regardless of where this family originated, it is quite successful in that it now occupies both continents of the New World as well as northern Australia, Asia, and Africa in the Old World. Globally, these frogs exhibit a variety of reproductive modes, including many that have direct development in which the tadpole stage takes place within the gelatinous coat of the eggs, which are placed in a terrestrial nest. However, all New World forms are characterized by small triangular heads, a fold of skin behind the eyes that can be drawn over the eyes when the frogs burrow, globose bodies with short hind legs (a toad-like body), and relatively small size. North American forms have aquatic larvae that are placed in exceptionally ephemeral pools of water and that grow and transform rapidly. North American microhylids belong to two genera, involving four species. One species is found in Alabama.

North American Narrow-mouthed Toads
Genus *Gastrophryne* (Fitzinger, 1843)

This is the genus of the Sheep Frogs or Narrow-mouthed Toads, so named because of the nasal call given by males and the small, triangular heads relative to the rest of the body. It is the sister genus to *Hypopachus*, a termite-eating frog of Central America (de Sá et al. 2012). These genera share a history with ten South American genera, suggesting an origin of New World members of the family Microhylidae in South America and recent invasion of North America. Members of *Gastrophryne* occur from the eastern United States to Honduras. Four species are included in the genus, one of which is found in Alabama.

Adult Eastern Narrow-mouthed Toad (*Gastrophryne carolinensis*), Bibb County, AL.

Eastern Narrow-mouthed Toad
Gastrophryne carolinensis (Holbrook, 1836)

DESCRIPTION Eastern Narrow-mouthed Toads are small frogs, attaining a maximum snout–vent length of about 1.4 in (35 mm). They have a tiny head, pointed snout, short hind legs (tibia length usually around 40 percent of snout–vent length), and smooth skin. The skin on the back of the neck has a transverse fold located behind the eyes, a handy field character for confirming the identity of this species. The hind toes are free of webbing, and each tympanum is concealed by a fold of skin. The dorsal ground color is brown, yellowish brown, or gray, usually with a large, dark figure oriented longitudinally along the middle of the back. The venter is dark gray with large white spots. Because of its short hind legs and globose body, members of this genus can only

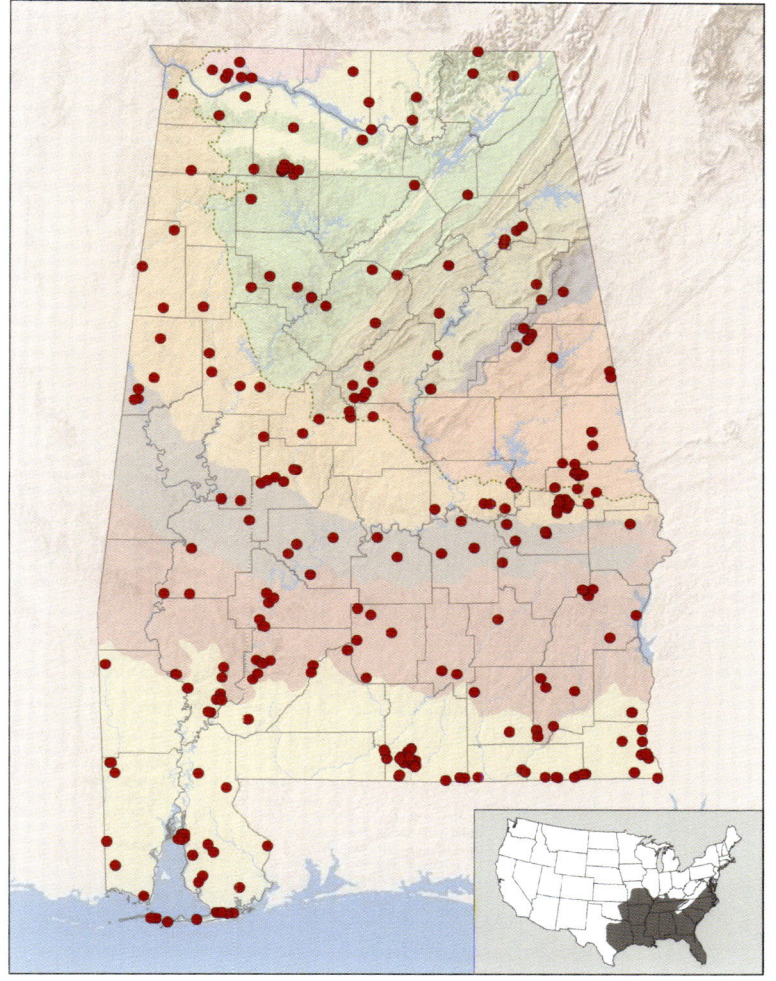

Distribution of Eastern Narrow-mouthed Toad (*Gastrophryne carolinensis*). Solid dots indicate localities of specimens or photographs examined by the authors or ADCNR/Natural Heritage Program occurrence records believed to be valid. Inset map depicts approximate range in the United States.

be confused with members of the genus *Anaxyrus*, which possess a pair or heel tubercles, and members of the genus *Scaphiopus*, which possess a single heel tubercle (tubercles absent in *Gastrophryne*).

ALABAMA DISTRIBUTION This species is abundant throughout the state.

HABITS The Eastern Narrow-mouthed Toad is secretive and spends most of its time in subterranean burrows, in decaying logs and stumps, or under rocks. A wide variety of habitats is used by this species, ranging from open pine forests to closed-canopy hardwood forests. Additionally, this species survives well in urban settings, especially where loose soils and thick organic layers are present. When exposed, these frogs may hop away quickly or attempt to burrow into rotting wood or debris. Activities are reduced during November through March when individuals remain in burrows.

Breeding activity in Alabama occurs from April to September and is elicited by heavy rains. Adults do not breed every year and move up to about 0.6 mi (1 km) to breeding sites (Dodd and Cade 1998). Breeding sites may include lakes, ponds, sloughs, and flooded roadside ditches, but ephemeral wetlands are the primary reproductive sites. Calling males are typically well hidden beneath clumps of grass or debris at the water's edge and are surprisingly difficult to locate. The call is a nasal, sheep-like bleat. Calling begins in the afternoon and continues into early evening and occurs sporadically throughout the breeding season (Bridges and Dorcas 2000). While in amplexus, which is pectoral in position, the male may become attached to the female's back by an adhesive substance secreted by special glands on his chest (Conaway and Metter 1967). Eggs are laid in groups of 10–90 as a surface film that is one egg-layer deep (Wright and Wright 1949).

Adults and juveniles migrate away from the breeding site where feeding and growth takes place. Diet items selected by Eastern Narrow-mouthed Toads are ants, almost exclusively. These ants tend to be species that are active at night, suggesting nocturnal foraging. Additionally, many ant species are consumed, including imported red fire ants, indicating that the anuran predators can forage in sites disturbed by human activities. Many of the ant species produce noxious chemicals that may be sequestered by the predators (Deyrup et al. 2013). Tadpoles are filter feeders, using movements of the mouth parts to intake water from which microscopic organisms are removed.

Adults produce skin toxins that are potent enough to kill mice when injected into the rodents and that cause vertebrate predators (Plain-bellied Watersnakes [Nerodia erythrogaster], Eastern Garter Snakes [Thamnophis sirtalis], Eastern Snapping Turtles [Chelydra serpentina], and Yellow-crowned Night Herons [Nyctanassa violacea]) to spit out these frogs when the predators bite them during predatory strikes (Garton and Mushinsky 1979). Tadpoles are mildly unpalatable to predators (Adams et al. 2011).

CONSERVATION AND MANAGEMENT This frog is exceptionally common and widespread. In fact, the species exhibits such low genetic variation across its entire geographic range that it appears to have experienced a recent genetic bottleneck followed by rapid range expansion (Makowsky et al. 2009). Because it is so common, Eastern Narrow-mouthed Toads receive no special protection from state law. The species can maintain apparently viable populations in urban garden areas that retain trees and deep litter (Guzy et al. 2012). Additionally, the species thrives in agricultural wetlands (Babbitt and Tanner 2000). However, environmental changes do affect this species. Selenium and strontium, contaminants from coal fly ash, can be transferred from a female Eastern Narrow-mouthed Toad to her offspring during egg production, causing reduced survival and increased deformities in frogs breeding at contaminated sites (Hopkins et al. 2006). Additionally, the species is known to carry ranaviruses, but at low prevalence and with low susceptibility (Hoverman et al. 2011).

TAXONOMY Despite its wide geographic distribution, this species has no morphological or genetic subspecific variation (Makowsky et al. 2008).

Ranid Frogs

Family Ranidae

The family Ranidae is distinguished from other frog families in the state of Alabama by the large size of its members, their lack of webbing on the front feet but extensive webbing on the hind feet, their lack of expanded toe pads, and the presence of smooth skin on the venter. This family is found in both the New and Old Worlds; contains more than 380 species, representing at least 25 genera; and has its greatest diversity in Asia, where it likely originated. This is the sister family to Mantellidae and Rhacophoridae, two other species-rich lineages of African, Asian, and Madagascan frogs (Pyron and Weins 2011). These frogs tend to live in or near water as adults, deposit eggs in an aquatic environment, pass through an aquatic tadpole stage, and transform to a juvenile form. Because of their large size and abundance in nature, members of this family frequently are the model species used in fields of study, such as physiology, to represent the entire group Anura. Additionally, ranid frogs are eaten by humans throughout the geographic range of the family. Two genera of this family occur within the United States, one of which occurs in Alabama.

American Water Frogs
Genus *Lithobates* (Fitzinger, 1843)

The genus *Lithobates* is New World in distribution and contains 50 species, 23 of which are found in the United States, with 10 occurring in Alabama. These are large frogs with heavily webbed feet that are almost always found in or near water. All have males that produce advertisement calls used to attract females. Males display pectoral amplexus, and females lay eggs in aquatic sites that are fertilized externally by the male. Females may deposit eggs as a film that floats on the surface of the water or create a globular mass of eggs that are carefully attached to aquatic vegetation. Once the eggs are deposited, the parents leave the eggs to hatch, grow throughout the tadpole stage, and transform into a juvenile frog. Adult *Lithobates* generally live near permanent water and have limited need to migrate from this center of activity. However, juveniles may disperse long distances over land to reach new wetland sites.

Our use of *Lithobates* is based on the resurrection of this genus by Frost et al. (2006). This treatment carved *Lithobates* out of the traditional concept of the genus *Rana*, which until recently included a paraphyletic assemblage of frogs from both the Old and New Worlds. Placement of the New World lineage of ranids into its own genus makes biogeographic sense to us as it highlights the single event that separated this lineage from ancestral lineages in the Old World and western United States. However, a second classification scheme for these frogs retains them in the genus *Rana* and identifies 16 subgenera for New World species (Hillis and Wilcox 2005; Yuan et al. 2016). In this scheme the subgenus *Lithobates* is restricted to five species from Central America, and Alabama ranids are placed in the subgenera *Scurrilirana* (Southern Leopard Frog), *Nenirana* (Crawfish, Dusky Gopher, Gopher, and Pickerel Frog), *Aquarana* (American Bullfrog, Pig Frog, Green Frog, and River Frog), and *Novirana* (Wood Frog; Hillis and Wilcox 2005; unassigned in Yuan et al. 2016).

Key to the Species of *Lithobates* of Alabama

1a Adults with dorsolateral ridges along dorsum; tadpoles not extremely large; **go to 2.**

Dorsal view of head, forelimbs, and body of Pickerel Frog (*Lithobates palustris*) showing dorsolateral ridges.

1b Adults without dorsolateral ridges along dorsum; tadpoles extremely large; **go to 7.**

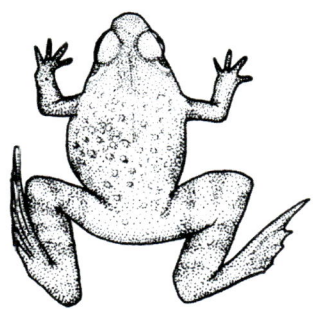

Dorsal view of head, forelimbs, and body of American Bullfrog (*Lithobates catesbeianus*) showing lack of dorsolateral ridges.

2a Face of adults with a wide dark mask extending from snout to behind tympanum; tadpole with a clear, tall tail.

Lithobates sylvaticus—**Wood Frog . . . page 93.**

From left to right:

Lateral view of head of Wood Frog (*Lithobates sylvaticus*).

Lateral view of Wood Frog (*Lithobates sylvaticus*) tadpole.

 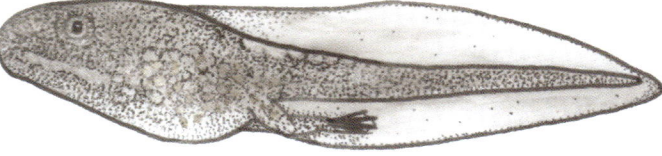

2b Face of adults may have dark stripe through eye but not in the form of a dark mask extending from snout to behind tympanum; tadpole not marked and shaped as above; **go to 3**.

3a Adults with two rows of large squarish spots along middle of back, some of these occasionally fused; tadpole with a tall, purplish fin lacking dark spots but possessing numerous white spots that do not align.

Lithobates palustris—**Pickerel Frog** . . . **page 98**.

From left to right:

Dorsal view of Pickerel Frog (*Lithobates palustris*).

Lateral view of Pickerel Frog (*Lithobates palustris*) tadpole.

3b Adults without spots or, if spotted, then spots neither squarish nor paired along middle of back; tadpole not colored as above; **go to 4**.

4a Dorsum of adults without numerous large, rounded dark spots; tadpole brown or gray.

Lithobates clamitans—**Green Frog** . . . **page 101**.

From left to right:

Dorsal view of Green Frog (*Lithobates clamitans*).

Lateral view of Green Frog (*Lithobates clamitans*) tadpole.

4b Dorsum of adults with numerous large, rounded dark spots (see illustrations with couplets 5–7); tadpole green (see illustrations with couplets 5–7); **go to 5.**

5a Adult body form not noticeably stubby; skin smooth and dorsal spots large; tadpole dark green with dark gray or black mottling on tail.

Lithobates sphenocephalus—**Southern Leopard Frog . . . page 105.**

From left to right:

Dorsal view of Coastal Plains Leopard Frog (*Lithobates sphenocephalus utricularius*).

Lateral view of Coastal Plains Leopard Frog (*Lithobates sphenocephalus utricularius*) tadpole.

5b Adult body form noticeably stubby; skin warty and dorsal spots small; tadpole light green with diffuse gray markings on tail; **go to 6.**

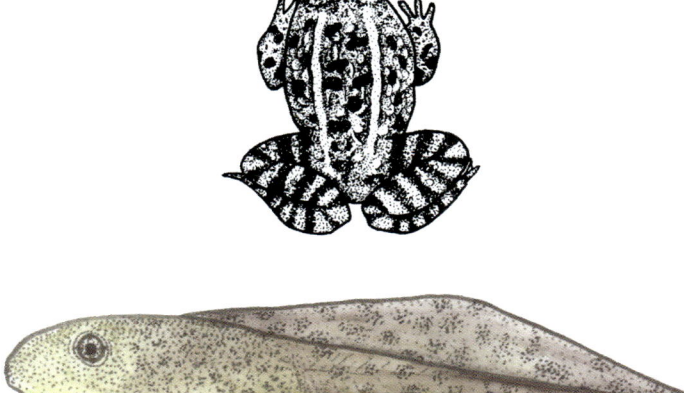

From left to right:

Dorsal view of Gopher Frog (*Lithobates capito*).

Lateral view of Gopher Frog (*Lithobates capito*) tadpole.

6a Mid-venter of chin with dark spots; dark dorsal spots lacking light border; dorsolateral ridge of males reddish.

Lithobates capito—**Gopher Frog** . . . page III.

Lithobates sevosus—**Dusky Gopher Frog** . . . page 116.

From left to right:

Ventral view of Gopher Frog (*Lithobates capito*) chin.

Dorsal view of Gopher Frog (*Lithobates capito*).

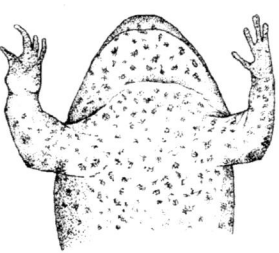
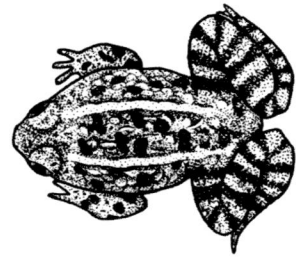

6b Mid-venter of chin immaculate; dark dorsal spots with light border; dorsolateral ridge of males yellow.

From left to right:

Ventral view of Northern Crawfish Frog (*Lithobates areolatus circulosus*) chin.

Dorsal view of Northern Crawfish Frog (*Lithobates areolatus circulosus*).

Lithobates areolatus circulosus—**Northern Crawfish Frog** . . . page 121.

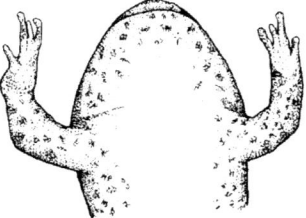
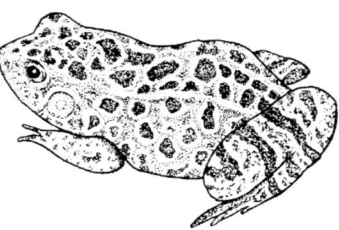

7a Venter of adults gray to gray-brown with light markings; tadpole gray or black and with a golden bar behind eye when small, tail clear with a smoky gray-to-black border fin and gray-to-black dorsal border to musculature.

Lithobates heckscheri—**River Frog** . . . page 125.

From left to right:

Ventral view of River Frog (*Lithobates heckscheri*).

Lateral view of River Frog (*Lithobates heckscheri*) tadpole.

7b Venter of adults white or whitish with dark markings; tadpole not as above (see illustrations with couplet 8); **go to 8.**

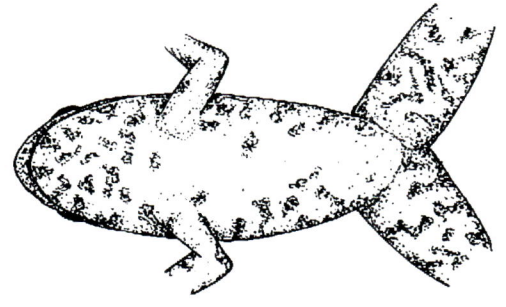

Ventral view of American Bullfrog (*Lithobates catesbeianus*).

8a Hind foot of adults with webbing extending well beyond terminal phalangeal joint to near tip of longest toe; rear of thigh usually with a light longitudinal stripe or with a longitudinal series of light spots; snout of adults pointed; tadpole brown with a black throat and caudal fin with aligned dark spots.

   Lithobates grylio—**Pig Frog . . . page 128.**

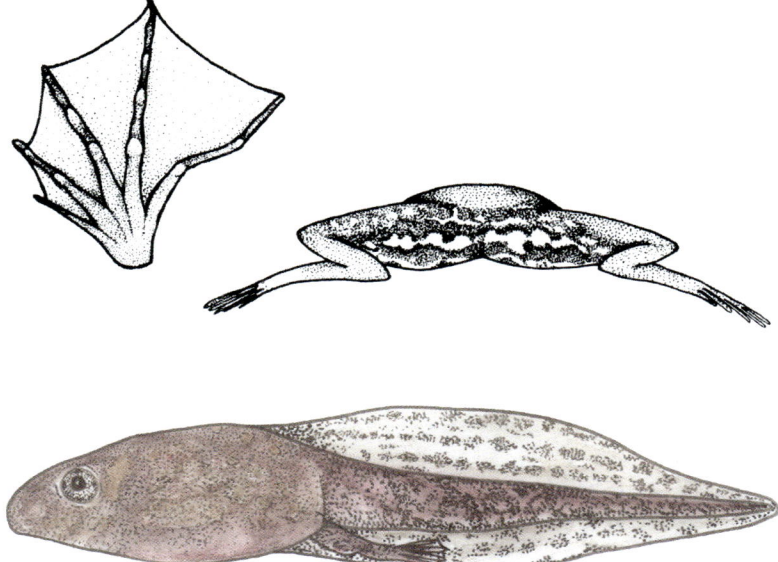

Clockwise from upper left:

Ventral view of hind foot of Pig Frog (*Lithobates grylio*).

Posterior view of thigh of Pig Frog (*Lithobates grylio*).

Lateral view of Pig Frog (*Lithobates grylio*) tadpole.

8b Hind foot of adults with webbing extending to terminal phalangeal joint of longest toe; rear of thigh without light longitudinal stripe or suggestion thereof; snout of adult rounded; tadpole green with unaligned black spots on body and tail (except for unspotted tadpoles from extreme southeastern Houston County).

Lithobates catesbeianus—**American Bullfrog** . . . **page 132.**

Clockwise from upper left:

Ventral view of hind foot of American Bullfrog (*Lithobates catesbeianus*).

Posterior view of thigh of American Bullfrog (*Lithobates catesbeianus*).

Lateral view of American Bullfrog (*Lithobates catesbeianus*) tadpole.

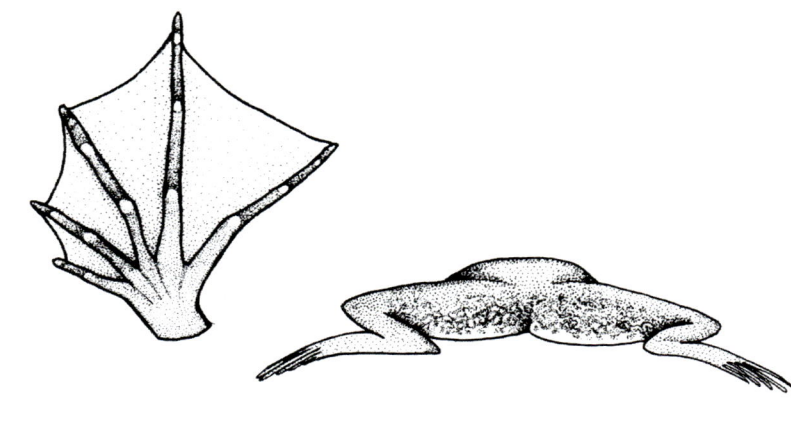

Adult Wood Frog
(*Lithobates sylvaticus*), Clay County, AL.

Wood Frog
Lithobates sylvaticus (Le Conte, 1825)

DESCRIPTION Wood Frogs are medium-sized ranids that attain a maximum snout–vent length of about 3.3 in (85 mm). The hind feet are extensively webbed, and the tips of the digits are not expanded. A pair of dorsolateral folds is present on the dorsum, and in calling males, a pair of vocal sacs expand dorsolaterally from the sides of the throat. The dorsal ground color is tan to brown, disrupted by a dark brown or black mask that extends from the tip of the snout, through the eye, to behind the tympanum. This dark mask is offset ventrally by a light stripe located just above the upper lip. The back and sides of the body may be uniform or may have scattered dark markings. The venter is uniform white or white with dark mottling under the throat and breast. The undersides of the femurs often are tinged with yellow. The tadpole is dark gray in color, has a diffuse light stripe along the lateral surface of the face, and has a light gray, spotless tail.

Wood Frogs are most like Green Frogs (*L. clamitans*) in general color pattern, but that species lacks a dark mask through the eye. The only other ranid from Alabama with a gray, spotless tadpole is the River Frog (*L. heckscheri*), but tadpoles of that species are much larger, have a dark smoky border to the tail fins, and are social.

ALABAMA DISTRIBUTION Wood Frogs are adapted to cool northern climates. As a result, this species is found in Alabama only within the

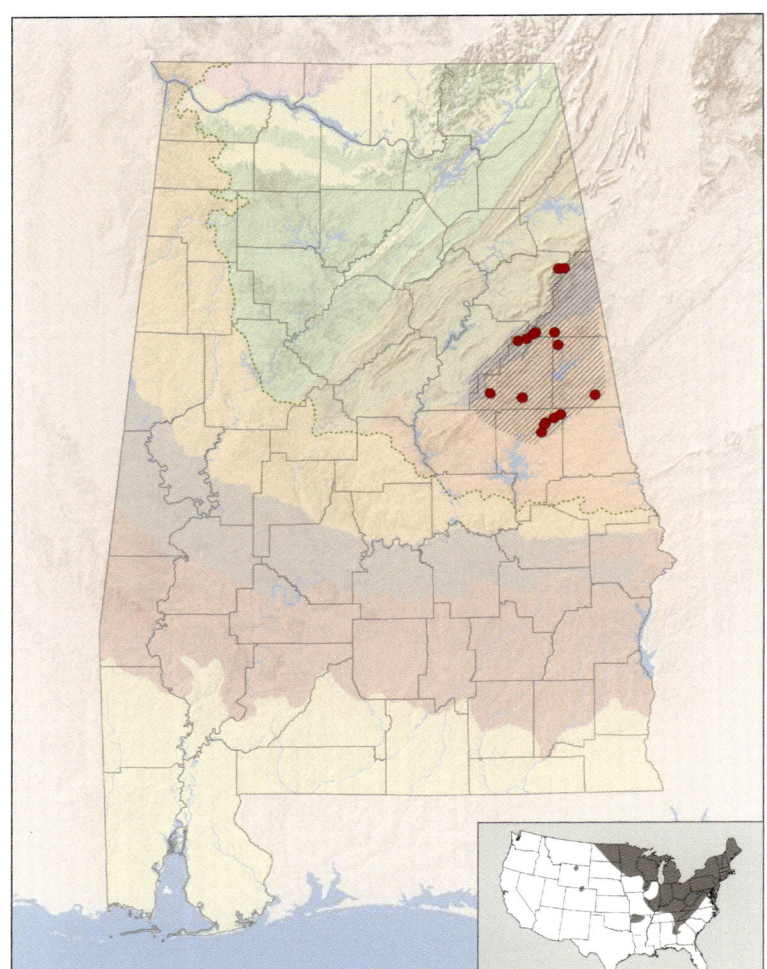

Distribution of Wood Frog (*Lithobates sylvaticus*). The presumed range of the species in Alabama is indicated by hatching. Solid dots indicate localities of specimens or photographs examined by the authors or ADCNR/Natural Heritage Program occurrence records believed to be valid. Inset map depicts approximate range in the United States.

Talladega Upland and Piedmont regions between 600 and 2,400 ft (180–730 m) elevation. Alabama populations are contiguous with populations of Wood Frogs in similar geological formations in the northern Piedmont of Georgia.

HABITS The Wood Frog is a terrestrial species, inhabiting mesic woodlands and the edges of pond and stream wetlands. By day, individuals seek refuge under rotting logs, leaf litter, and rocks, emerging at night to feed. The color of this species camouflages it well on a leafy forest floor so that, when active, it is not likely to be noticed unless it jumps.

Breeding in Alabama occurs during January or February when males and females migrate to enter vernal pool wetlands. Breeding is explosive, with all activity taking place over a period of a few days during each year. Adults migrate to the breeding sites along moist forested corridors (Vasconcelos and Calhoun 2004). About 18 percent of frogs settle at a breeding pond that differs from the birth pond, but once an adult breeds at a particular pond, the frog will remain faithful to the pond (Berven 1990a). The mating call of males has been described as a rasping "craw-aw-auk," suggesting the quacking of a duck, and has limited volume. Females enter the breeding site, apparently following the vocal cue of the males. However, the sex ratio of breeding aggregations is heavily male biased (Berven 1990b), and there is intense male-male competition to amplex each female. Waves created by frogs moving in the water are detected by males and cause them to approach the source of the waves (Höbel and Kolodziej 2013). If the individual moving is a female, several males will approach and grasp her, with the male possessing the strongest grip being the one that is likely to gain the position that is most likely to fertilize the eggs; this set of features results in mating in which the largest males in the population tend to be found in amplexus with the largest females entering the breeding pond (Howard and Kluge 1985). Females deposit up to 2,500 eggs underwater and attach them to aquatic vegetation or lay them unattached in shallow water. Nests are placed next to those of other clutches, creating large communal egg masses; eggs toward the center of such masses have higher survival rates because they are buffered from extreme temperatures (Waldman 1982). The tadpoles that emerge from these eggs are known to be palatable to fish predators (Kats et al. 1988).

After breeding, adult frogs move back to forested habitats where more feeding occurs. Adults are predatory, primarily consuming insects and other invertebrates but can eat small vertebrates. Tadpoles eat periphyton and phytoplankton that are scraped from aquatic vegetation. Tadpoles may also scavenge nutrients for dead tadpoles. Tadpoles transform after 2–5 months of larval growth and migrate to woodlands to overwinter.

CONSERVATION AND MANAGEMENT Wood Frogs occur as isolated populations in Alabama that, along with sites in adjacent Georgia, represent the southern extent of its geographic range. Therefore, conservation efforts for this species will largely come from large populations located elsewhere. The species is listed as imperiled by the ANHP and is listed as Priority 2 (species of high conservation concern) by ADCNR (Shelton-Nix 2017). Populations in the Talladega and Shoal Creek Districts of Talladega National Forest, Cheaha State Park, and Horseshoe Bend National Military Park are key conservation lands for Wood Frogs within Alabama.

Several features of the population biology suggest that maintenance of Alabama's population will be challenging. Adult survival is positively associated with mean monthly rainfall (Berven 1990a), so climate change might extirpate the species from the state if such change involves a reduction in rainfall. Additionally, population size in this species fluctuates widely among years, largely because of periods of complete failure of tadpoles to metamorphose followed by years of high juvenile recruitment (Berven 1995). Therefore, retention of the species in Alabama also will require maintaining a metapopulation structure that will allow rescue via dispersal to populations that might become extirpated. However, Petranka et al. (2007) demonstrated that carefully designed artificial ponds can assist population persistence. Long-term data suggest that calling in these frogs is occurring progressively earlier, perhaps caused by climate change (Walpole et al. 2012). Loss of populations along the southern extent of the geographic range of this species might be expected if warming climates force these cold-adapted frogs northward.

Maintenance of breeding sites is crucial to management for this species. These sites must remain fish-free, and forested habitat around these sites must remain wooded for at least 109 yd (100 m) to keep them suitable for seasonal migratory activities (Freidenfelds et al. 2011). Logging during winter near breeding ponds is likely to extirpate populations of adults, whereas such destruction during other times of the year likely will not cause extirpation because adults move to upland habitats (Regosin et al. 2003). However, if logging does occur, then retention of brush piles likely increases survival of remaining frogs by decreasing desiccation (Rittenhouse et al. 2008). Timber harvests that create open gap areas are avoided by adults and metamorphs of this species (Strojny and Hunter 2010).

The tadpole stage of Wood Frogs is sensitive to numerous disease and chemical stressors. This stage can experience mass mortality due to infection with chytrid fungus (*Batrachochytrium dendrobatidis*), and the spread of this disease may be enhanced by transfer from bullfrogs, which typically are not affected by the disease (Greenspan et al. 2012). Similarly, mass mortality of tadpoles and juveniles in association with high prevalence of ranavirus is known for these frogs (Miller et al. 2011). Tadpole growth and survival are not affected by typical doses of the herbicide glyphosate (Lanctôt et al. 2013), but prolonged exposure to this chemical reduces developmental rates (Navarro-Martín et al. 2014). Exposure to the fungicide triphenyltin can be lethal to tadpoles (Higley et al. 2013). Proximity to roads is known to increase skeletal abnormalities in the development of Wood Frogs (Reeves et al. 2008), perhaps because of the runoff of toxic chemicals. Finally, metamorphosis is slowed in wetlands created to mediate oil extraction (Hersikorns and Smits 2011).

TAXONOMY Wood Frogs represent the basal member of the genus *Lithobates*, suggesting that this lineage originated in cold northern climates, expanding to occupy southern tropical climates (Yuan et al. 2016). Despite having the largest geographic distribution of any North American anuran, the species is considered to have no subspecific variation.

Adult Pickerel Frog (*Lithobates palustris*), Jackson County, AL.

Pickerel Frog
Lithobates palustris (Le Conte, 1825)

DESCRIPTION This is a medium-sized ranid frog, attaining a maximum snout–vent length of about 3.1 in (80 mm). The hind foot has extensive webbing between the toes, and the toes lack expanded pads. The tympanum is about the same size as the eye, and the dorsum possesses a pair of dorsolateral folds. In color, these frogs have a grayish dorsum with 8–14 bold, squarish dark spots that are arranged in pairs along two rows between the folds. On occasion, adjacent paired spots may be fused. A small spot is present above each eye, and often one is present on the snout. The sides have scattered dark spots, the dorsal surfaces of the legs are barred, and the inner surfaces of the thighs, along with the posterior portion of the venter, are suffused with yellow. During the tadpole stage, the body and tail musculature have a purple cast and the tail fin is covered with small dark spots.

This species is similar in shape and general appearance to the Southern Leopard Frog (*L. sphenocephalus*), but that species has round dark spots between the folds that do not align as pairs. No other ranid tadpole in Alabama has the purplish cast of Pickerel Frogs.

ALABAMA DISTRIBUTION Pickerel Frog occurrences are concentrated on the steep slopes of the Talladega Upland, the Ridge and Valley, the Sequatchie Valley, the mountains of Jackson and Madison Counties, and the walls of the Moulton and Tennessee River Valleys. Disjunct Coastal Plain populations are found in steeply sloped regions of the Red Hills (Marengo, Monroe, and Wilcox Counties), and the species is recorded from a cave near Brooklyn, Conecuh County, at the upper edge of the Lower Coastal Plain.

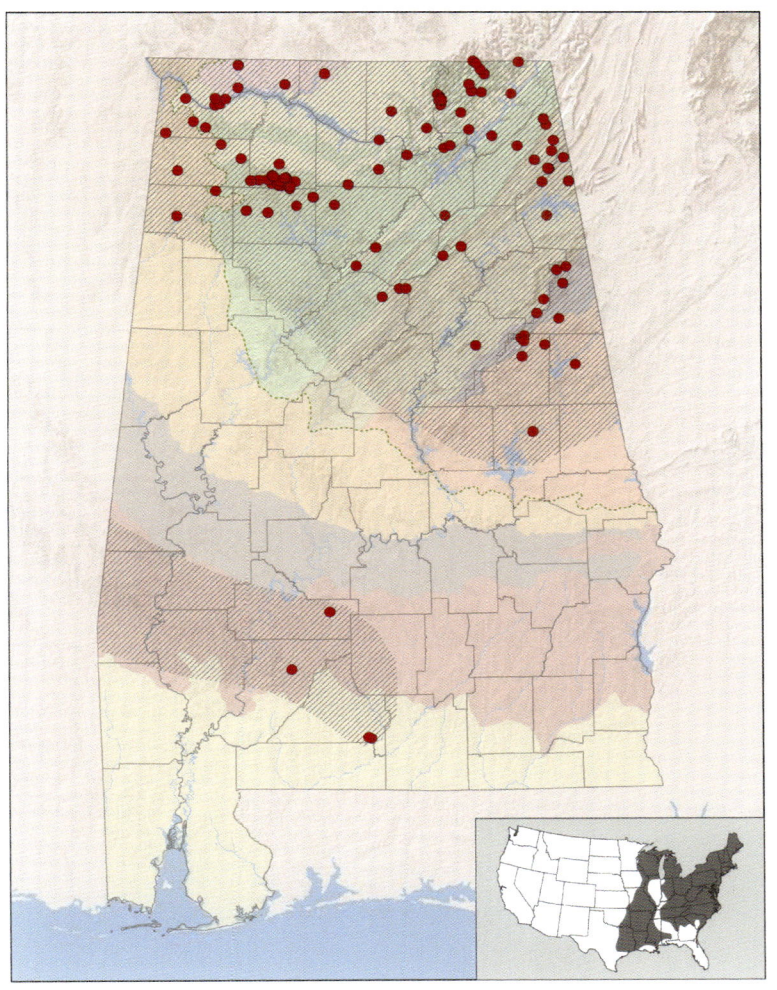

Distribution of
Pickerel Frog
(*Lithobates palus-*
tris). The presumed
range of the spe-
cies in Alabama is
indicated by hatch-
ing. Solid dots
indicate localities
of specimens or
photographs exam-
ined by the authors
or ADCNR/Natural
Heritage Program
occurrence records
believed to be
valid. Inset map
depicts approxi-
mate range in the
United States.

HABITS In Alabama, as well as throughout most of its range, the Pickerel Frog tends to be associated with cool, clear water in ravine forests
and meadows. Breeding occurs during winter and early spring when
males attract females via a call that is a variable, low-pitched croaking
sound that often has a snore-like quality. Males may call throughout
all nighttime hours and tend not to call during daylight, although calling varies dramatically among sites (Todd et al. 2003). Eggs are laid
in globular masses of up to about 3,000 per mass, usually in a woodland pool or a quiet pocket in a small stream. Tadpoles may overwinter,

transforming to the adult form the following activity season (Wilbur and Fauth 1990). Adults remain in the general area of the breeding site throughout the year, where they feed in forested areas and burrow in mud of aquatic habitats to avoid temperature extremes. However, populations near caves may migrate en masse to warm areas in caves during winter months where they remain active all year (Resetarits 1986).

Adult frogs are predatory, primarily consuming insects and other invertebrates, but can eat small vertebrates. Tadpoles eat periphyton and phytoplankton that are scraped from aquatic vegetation. Tadpoles may also scavenge nutrients from dead tadpoles.

Pickerel Frogs produce skin secretions that are toxic to many reptiles and other amphibians and, therefore, should not be confined with other species. Additionally, care should be taken not to rub these secretions in one's eyes to avoid the stinging sensation that will occur. Bradykinin-related peptides are known from these skin secretions (McCrudden et al. 2007) and may stimulate pain receptors and cause spasms in smooth muscles, leading to regurgitation, in predators.

CONSERVATION AND MANAGEMENT Pickerel Frogs receive no regulatory protection in Alabama. The species has a relatively wide distribution in forested ravines of the Talladega Upland, upper Ridge and Valley, and Appalachian Plateaus. Because it lives in steep topography where farming and development generally has not occurred, apparently viable populations are common, but population size is not exceptionally high in any of them. Retention of forested riparian zones and meadows along streams are vital for maintaining healthy populations of this species. Land owners desiring to enhance habitats for Pickerel Frogs should avoid any activities that alter siltation and produce chemical runoff of herbicides or insecticides. Additionally, road noise is known to reduce habitat occupancy by this species (Cosentino et al. 2014). The species is known to be susceptible to ranaviruses, a disease that may cause mass mortality (Hoverman et al. 2011; Miller et al. 2011). Long-term data from the northeastern US documents the decline of occupancy of this species over time (Weir et al. 2014).

TAXONOMY This species is the basal member of the *areolatus* subgroup (*L. areolatus, L. capito, L. palustris,* and *L. sevosus*) of leopard frogs (*pipiens* group; Hillis and Wilcox 2005). It is considered to have no subspecific variation.

Adult Green Frog (*Lithobates clamitans*), Mobile County, AL.

Green Frog
Lithobates clamitans (Latreille, 1801)

DESCRIPTION Green Frogs are medium-sized ranids, attaining a maximum snout–vent length of about 3.9 in (100 mm). The hind foot has extensive webbing between the toes, which are pointed and lack expanded toe pads. The diameter of the tympanum is about the same size as the eye, and there is a pair of dorsolateral folds on the dorsum. The skin of the dorsum may be rough in texture. The dorsal ground color of the head of adults frequently is bright green, and the body typically is greenish brown or bronze and generally lacks distinct spots in the southern part of the state and possess dark spots in the northern part. The edge of the jaw has alternating dark and light spots. On the venter, a white ground color is disrupted by wormlike dark markings, which are sometimes confined to the legs and throat. The throat of males may be bright yellow. The tadpole is light greenish gray, has tail musculature that is light gray with diffuse dark spots, and possesses tail fins with diffuse dark spotting.

Adult Green Frogs are most like American Bullfrogs (*L. catesbeianus*), Pig Frogs (*L. grylio*), and River Frogs (*L. heckscheri*), but those three species lack dorsolateral folds. Tadpoles of Green Frogs are most like those of Southern Leopard Frogs (*L. sphenocephalus*), but that species has darker spotting along the tail.

ALABAMA DISTRIBUTION Green Frogs are found throughout Alabama.

HABITS Swamps, floodplain pools, and small streams are the favored habitats of this common frog, although it may occur in other types

Adult Green Frog (*Lithobates clamitans*), Bibb County, AL.

of aquatic situations. These frogs tend to be less wary than Southern Leopard Frogs, another Alabama ranid of about the same size, and are thus more easily captured. The home range occupied by these frogs is also where reproduction takes place, requiring no migration to a breeding site. Breeding of Green Frogs in Alabama begins as early as April and lasts into August or September, with peak mating during May, June, and July. The call note of males is an explosive "clung." The call is issued from one to three times per calling bout, and calls are produced mostly at night with peak vocalizations from midnight to dawn; males also tend to call persistently throughout the breeding season (Bridges and Dorcas 2000), peaking on warm nights in spring (Steen et al. 2013). Males are territorial, defending areas of the aquatic margin from invasion by other males. Defensive behaviors include calling and splashing, as well as physical confrontation of intruders (Wells 1978). Neighboring males differ from each other in call structure, and this allows resident males to distinguish nearest neighbors, with which the resident has already settled border disputes, from novel intruders (Bee et al. 2001). Females are attracted to the calls of males but also choose a mate based on the quality of his territory, which also serves as the nest site for the female (Wells 1977). Once a female selects a male, he amplexes her, and she eventually lays several thousand eggs as a surface film that is fertilized externally by the male. Tadpoles of this species are known to be unpalatable to a variety of aquatic predators (Kats et al. 1988; Adams et al. 2011), allowing them to overwinter in the tadpole stage and transform during warm spring months. Adults remain in water or on moist soil under leaf litter at the breeding site. Thus, the

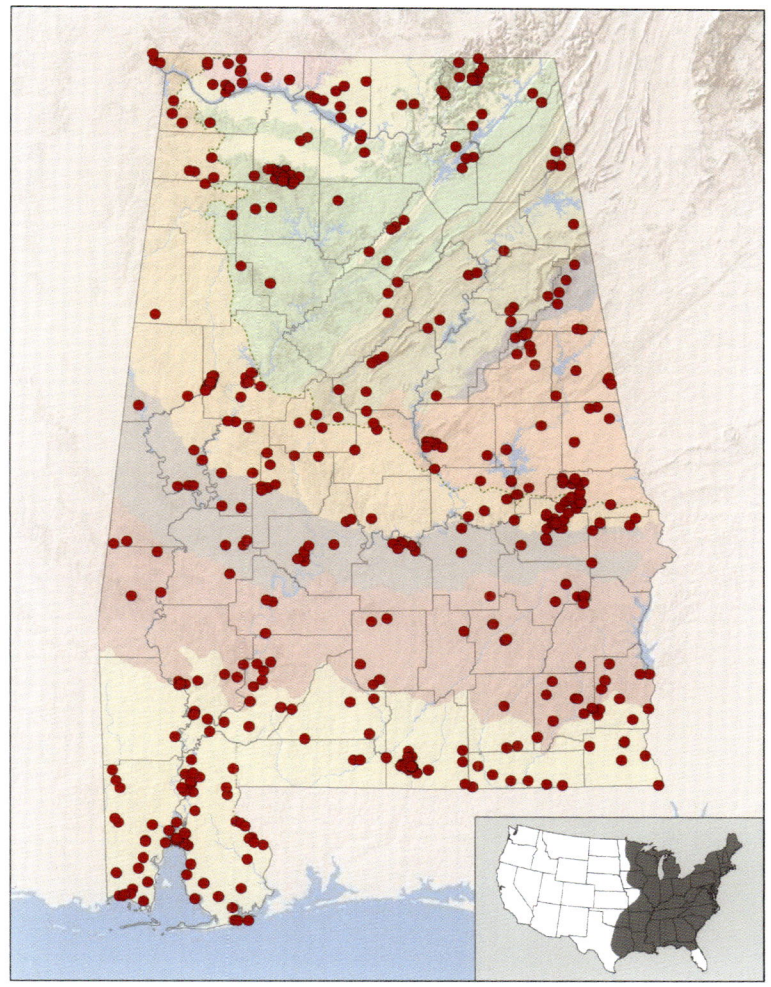

Distribution of Green Frog (*Lithobates clamitans*). Solid dots indicate localities of specimens or photographs examined by the authors or ADCNR/Natural Heritage Program occurrence records believed to be valid. Inset map depicts approximate range in the United States.

species typically does not migrate to breeding sites, but these frogs may move long distances along streams during fall months.

Adult frogs are predatory, primarily consuming insects and other invertebrates, but can eat small vertebrates. Tadpoles eat periphyton and phytoplankton that are scraped from aquatic vegetation. Tadpoles may also scavenge nutrients from dead tadpoles.

CONSERVATION AND MANAGEMENT This species receives no regulatory protection in Alabama. It is abundant, thriving in natural wetlands and streams as well as those that have been heavily modified by urbanization. Therefore, special management practices do not appear to be necessary to retain this species in Alabama's landscape. The species does not thrive in areas that have been clear-cut, unless brush piles are retained to decrease mortality associated with desiccation (Rittenhouse et al. 2008). Timber harvests that create open gap areas are avoided by metamorphs of this species (Strojny and Hunter 2010). Calling males respond to road noise by reducing their calls (Cosentino et al. 2014), but long-term calling phenology is not changed in the face of climate change (Walpole et al. 2012). Patch occupancy of Green Frogs increases if wetlands are surrounded by agricultural lands (Alix et al. 2014a). Use of the herbicide glyphosate in wetlands occupied by Green Frogs increases infection of tadpoles by chytrid fungus (*Batrachochytrium dendrobatidis*; Edge et al. 2011). Similarly use of the herbicide trifuralin and the insecticide carbaryl negatively affect development of tadpoles (Boone et al. 2013; Weir et al. 2012). Chytrid fungus is known to infect this species, but with limited negative effects. Thus, the species may serve as a carrier for the disease (Gahl et al. 2011). Mass mortality of tadpoles and juveniles is known in association with the presence of ranaviruses (Homan et al. 2013).

TAXONOMY Green Frogs are the sister species to the Florida Bog Frog (*L. okaloosae*) and form a component of the *catesbeianus* group of ranids native to eastern North America (Hillis and Wilcox 2005). Traditionally, two subspecies of this common species have been recognized, both of which occur in Alabama. Mount (1975) considered Bronze Frog (*L. c. clamitans*) to be restricted to the Lower Coastal Plain, Northern Green Frog (*L. c. melanota*) to be restricted to the northern part of the state, and the central portion of the state to contain a wide zone of intergradation between the two. Based on evidence from the mitochondrial genome, Austin and Zamudio (2008) also distinguished two putatively monophyletic clades, one for the extreme Gulf and Atlantic Coastal Plains of the southeast (Coastal Plains–Eastern Appalachian clade) and one for all other areas (Widespread clade). However, the distributions of the genetic lineages are not concordant with those of the traditional subspecies, and the characteristics used to distinguish these subspecies are subtle. Therefore, we follow Frost et al. (2017) in considering this to be a single species without subspecific variation. Because the mitochondrial clades occur sympatrically within Alabama and Georgia, we do not view them as representing cryptic taxa.

Southern Leopard Frog
Lithobates sphenocephalus (Cope, 1886)

TAXONOMY Hillis (1988) placed the Southern Leopard Frog in the species *Lithobates sphenocephalus* (considered to be the genus *Rana*; Pauly et al. 2009), separating this species of the southeastern United States from the Northern Leopard Frog (*L. pipiens*). That study showed the Southern Leopard Frog to be part of a widespread complex of spotted ranids, and subsequent analyses by Hillis and Wilcox (2005) showed this species to be nested within the *berlandieri* subgroup of leopard frogs. The *berlandieri* subgroup contains approximately 50 species distributed largely in Central America and western North America but invading the eastern United States (Yuan et al. 2016).

Two subspecies of Southern Leopard Frog typically have been recognized, one restricted to peninsular Florida (Florida Leopard Frog [*L. s. sphenocephalus*]), and the other widely distributed across the southeastern United States (Coastal Plains Leopard Frog [*L. s. utricularius*]). Based on analyses of the mitochondrial genome, Newman and Rissler (2011) documented two distinct lineages, a western clade covering most of Alabama and the rest of the region west of the Appalachian Mountains, and an eastern clade covering extreme southeastern Alabama and the region east of the Chattahoochee River and Appalachian Mountains. Because niche models generated by Newman and Rissler (2011) identified the eastern lineage as being centered in Florida, that lineage was interpreted to represent *L. sphenocephalus sphenocephalus* and the western lineage to represent *L. sphenocephalus utricularius*.

Because both lineages were documented from Alabama, we consider both to be present in the state but place them in a single account because we know of no morphological features to distinguish them in the field.

Adult Florida Leopard Frog (*Lithobates sphenocephalus sphenocephalus*), Covington County, AL.

Florida and Coastal Plains Leopard Frogs
Lithobates sphenocephalus sphenocephalus (Cope, 1886)
Lithobates sphenocephalus utricularius (Harlan, 1825)

DESCRIPTION Florida and Coastal Plains Leopard Frogs are medium-sized ranids, attaining a maximum snout–vent length of around 5.1 in (130 mm). The hind foot is extensively webbed, the tips of the digits are not expanded, and a pair of dorsolateral folds is present in these frogs. In calling males, a pair of vocal sacs expands laterally from the throat region. The dorsal ground color is green to brownish with elongate or rounded dark spots that do not align as pairs along the middle of the dorsum. Most spots have light borders, but in occasional forms (*burnsi* mutants) spots are completely lacking. The head usually is green, and the upper jaw has a distinct light line along the upper lip. The dorsal aspect of each tibia is marked with interrupted bars, and the backs of the thighs have a dark reticulum. The venter is uniform white, and each tympanum has a light spot in its center. These two subspecies are most easily confused with Pickerel Frog (*L. palustris*) but lack the paired square-shaped dark markings of that species. Individuals with the *burnsi* color pattern are easily confused with the bronze morph Green Frog (*L. clamitans*) but are more slender and have a narrower snout than that species. The tadpole has a dark green body with a tail that is covered in large dark spots that have diffuse borders.

Adult Coastal Plains Leopard Frog (*Lithobates sphenocephalus utricularius*), Macon County, AL.

These tadpoles are most like those of Crawfish Frogs (*L. areolatus*), Green Frogs, and Dusky Gopher Frogs (*L. sevosus*), but those species lack extensive dark spotting to the tail.

ALABAMA DISTRIBUTION In combination, these two subspecies are found throughout the state. We infer that those forms from Mobile Bay eastward in the Perdido, Yellow, Conecuh, Pea, Choctawhatchee, and Lower Chattahoochee drainages are Florida Leopard Frogs and those in the remainder of the state are Coastal Plains Leopard Frogs.

HABITS These subspecies occupy most types of aquatic habitats (Liner et al. 2008) during their annual cycle of activities. On rainy nights Florida and Coastal Plains Leopard Frogs are often seen on highways, and by day these subspecies can be found along wetland margins, typically within 2–3 leaps of an aquatic refuge. Adults migrate to breeding sites at night during and immediately after rain showers (Todd and Winne 2006). Preferred breeding sites include permanent and semi-permanent woodland pools, but breeding also occurs in flooded roadside ditches, ponds, lakes, and beaver swamps.

The breeding call is a series of guttural croaks and clucks. Smith (1961) aptly states that the call can be roughly simulated by rubbing a thumb across an inflated balloon. Florida and Coastal Plains Leopard Frogs may breed at any time of the year in Alabama when heavy rains coincide with temperatures above 10°C. However, most breeding occurs from December through March (Steen et al. 2013). Males may call during daylight and nighttime hours, and calling activity tends to

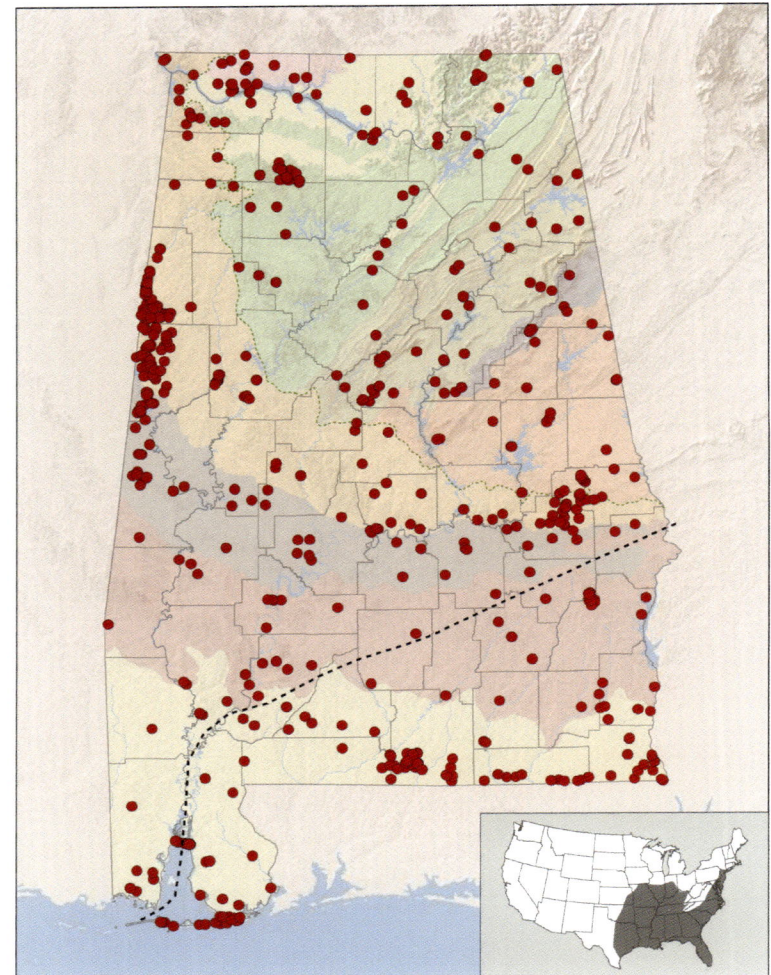

Distribution of Florida Leopard Frog (*Lithobates sphenocephalus sphenocephalus*) and Coastal Plains Leopard Frog (*Lithobates sphenocephalus utricularius*). Solid dots indicate localities of specimens or photographs examined by the authors or ADCNR/Natural Heritage Program occurrence records believed to be valid. The dashed line indicates a hypothesized boundary between the Florida Leopard Frog (southeastern segment of state) and Coastal Plains Leopard Frog (rest of state). Inset map depicts approximate range in the United States.

be similar among sites (Todd et al. 2003). However, calling occurs sporadically during the breeding season (Bridges and Dorcas 2000). The eggs are laid in shallow water in masses of a few hundred to several thousand per mass. Females of Florida and Coastal Plains Leopard Frogs lay their eggs as globular masses that frequently are attached below the water surface to stems of aquatic vegetation; however, masses may be deposited on the bottom surface without attachment to any structure. These two subspecies frequent the same breeding sites as Gopher Frogs and Dusky Gopher Frogs, which also lay eggs in similar places and at similar times of year. Eggs of the two species of Gopher Frog are slightly larger and have gelatinous outer coats that are firmer.

Adult frogs are predatory, primarily consuming insects and other invertebrates, but can eat small vertebrates. Tadpoles eat periphyton and phytoplankton that are scraped from aquatic vegetation. Tadpoles may also scavenge nutrients from dead tadpoles.

During the day Florida and Coastal Plains Leopard Frogs are frequently flushed from around the margins of aquatic sites. These two subspecies are alert, active, and strong jumpers, often leaping two to three times in rapid succession before landing and remaining motionless. Florida and Coastal Plains Leopard Frogs produce toxins that are potent and can kill other frogs if they are housed together. Additionally, if these toxins are rubbed into human eyes, they will cause stinging that will require flushing with copious amounts of water. Bradykinin-related peptides are known from these skin secretions (Sin et al. 2008) and may stimulate pain receptors and cause spasms in smooth muscles, leading to regurgitation, in predators. Tadpoles of these two subspecies are unpalatable to predators (Adams et al. 2011).

CONSERVATION AND MANAGEMENT These two abundant subspecies receive no regulatory protection in Alabama. In fact, these frogs do not appear to be affected by urbanization (Guzy et al. 2012). However, long-term data from the northeastern US suggest that patch occupancy of this species has declined over time (Weir et al. 2014).

Habitat quality is known to affect populations of Florida and Coastal Plains Leopard Frogs. The increased number of Coastal Plains Leopard Frogs captured on plots burned every year and every seven years, compared with unburned control plots, suggests that this species is adapted to fire-maintained pine savannas and that management with fire can increase the occurrence of this species (Halstead 2007). Coastal Plains Leopard Frogs declined at Rainbow Bay in South Carolina over a 35-year period of monitoring, either because of decreased rainfall (Daszak et al. 2005), or because of a lack of fire. Occupancy of Florida and Coastal Plains Leopard Frogs can be increased when reproductive wetlands are surrounded by agricultural lands (Alix et al. 2014a; Babbitt and Tanner 2000). Heavy road morality can occur (Smith and Dodd 2003). However, this source of mortality does not affect abundances of the two subspecies, and so no special management efforts are needed to reduce this source of mortality. Leaves of the invasive Chinese Tallow Tree (*Triadica sebifera*) lower pH and oxygen, leading to death of leopard frog eggs (Adams and Saenz 2012), but surviving tadpoles may metamorphose at a larger body size (Cotton et al. 2012).

Contaminants and disease agents also affect populations of leopard frogs. The pesticide carbaryl increases mortality and decreases size at metamorphosis for tadpoles of Coastal Plains Leopard Frogs (Bridges 2000). Endrin and Toxaphene cause increased mortality in tadpoles (Hall and Swineford 2003). Nitrate kills tadpoles and perchlorate inhibits metamorphosis (Ortiz-Santaliestra and Sparting 2007), whereas atrazine causes reduced algae growth, which decreases size and increases time to metamorphosis of Coastal Plains Leopard Frogs (Boone and James 2003). The presence of environmental copper reduces egg survival (Lance et al. 2012). Herbicides, insecticides, metals, and fertilizers must be used with caution if retention of leopard frogs is a management goal. However, use of the fungicide thiophanate-methyl enhances tadpole growth, perhaps by controlling pathogens such as chytrid fungi (Hanlon et al. 2012). Chytrid fungal disease (*Batrachochytrium dendrobatidis*) agents are known from leopard frogs (Pullen et al. 2010), and mass mortality of tadpoles and juveniles in association with high prevalence of ranavirus is known (Miller et al. 2011).

Adult Gopher Frog (*Lithobates capito*), Covington County, AL.

Gopher Frog
Lithobates capito (LeConte, 1855)

DESCRIPTION This is a rather large, stout-bodied frog, attaining a maximum snout–vent length of about 3.9 in (100 mm). The hind feet have extensive webbing between the toes, and the toes are pointed, lacking expanded toe pads. The diameter of the tympanum is similar to that of the eye, and dorsolateral ridges are present. In calling males, the vocal sacs are paired and emerge laterally from under the throat. The body is extremely rugose due to the presence of enlarged wart-like structures that may obscure the dorsolateral folds. In color, the dorsum is gray to light brown with dark gray blotches that are separated from smaller dark markings of varying shapes. The venter is light gray with numerous small dark gray spots, many of which coalesce. The groin and inner surfaces of the thighs are tinged with yellow in live animals. The tadpole is light green with diffuse dark mottling on the tail fins.

In morphology and color pattern adults and tadpoles of this species are indistinguishable from Dusky Gopher Frog (*L. sevosus*). However, the relatively short hind legs and warty body make adults of these two species easily distinguishable from all other ranids in Alabama except Northern Crawfish Frogs (*L. areolatus circulosus*), which lack spots on the center of the chin and have dark dorsal spots that are outlined by light gray or tan. Tadpoles of Northern Crawfish Frogs, Gopher Frogs, and Dusky Gopher Frogs are most like those of Southern Leopard

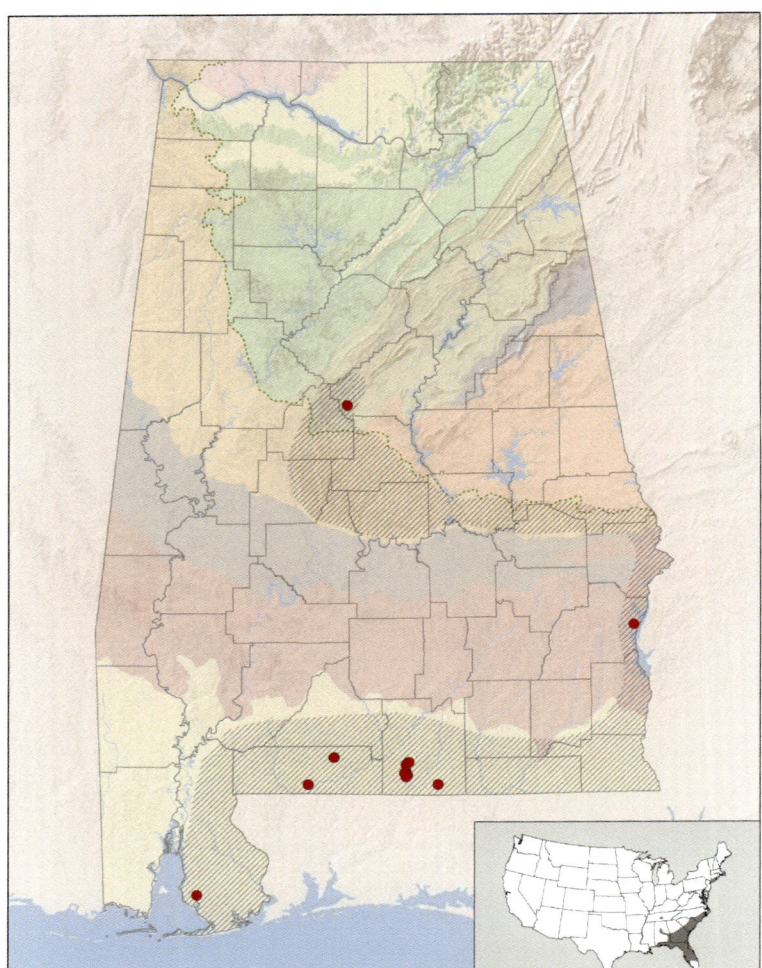

Distribution of Gopher Frog (*Lithobates capito*). The presumed range of the species in Alabama is indicated by hatching. Solid dots indicate localities of specimens or photographs examined by the authors or ADCNR/Natural Heritage Program occurrence records believed to be valid. Inset map depicts approximate range in the United States.

Frogs (*L. sphenocephalus*), but that species is darker green in body color and has darker mottling of the tail fin.

ALABAMA DISTRIBUTION Gopher Frog occurrences are clustered within the Coastal Plain but extend into sandy ridges within the Red Hills (Barbour and Coffee Counties) and sandy regions of the perimeter of the Ridge and Valley (Shelby County). The species appears to have been extirpated from sites in Barbour and Shelby Counties described by Mount (1975).

HABITS This shy frog inhabits burrows of Gopher Tortoises (*Gopherus polyphemus*), rodent burrows, and stump holes during most of the year (Roznik and Johnson 2009), where they apparently feed and do little else until the reproductive season. They emerge at night but remain in proximity to the refuge entrance. These sites are vacated during the reproductive season, when individuals migrate up to 3.1 mi (5 km) to breeding sites but are returned to persistently along consistent migration routes (Blihovde 2006; Humphries and Sisson 2012). Gopher Frogs breed during the winter months of October through May following heavy rains (Palis 1998). Males migrate to the breeding site earlier than do females and remain there two to three times longer than do females (Bailey 1991; Palis 1998). These breeding sites are fish-free, temporary, depressional wetlands or marshes (Bailey 1991; Liner et al. 2008). Sinkhole ponds are particularly favored reproductive sites. At these sites, males attract females with a deep, rumbling, snore-like advertisement call that frequently is given under water. When given above water, the call may be heard from a distance of at least 0.25 mi (0.4 km). Females remain at the reproductive site only long enough to mate and deposit eggs (approximately 9 days; Bailey 1991). Both sexes enter and exit the breeding site at the same general places, suggesting that dispersal corridors are important landscape features (Palis 1998). Eggs are deposited in a globular mass that is attached to stems of aquatic vegetation near the edge of the wetland. These masses contain about 2,000 eggs (Palis 1998) and are quite similar in appearance to those of Southern Leopard Frogs (*L. sphenocephalus*) but tend to be larger and have gelatinous outer coats that are firmer. The number of egg masses produced is positively correlated with average rainfall during the breeding season, perhaps because frogs can migrate from longer distances during years with increased rainfall (Jensen et al. 2003). Juvenile frogs migrate from the breeding site from May to August, and these migrations are not associated with rain (Greenberg 2001).

Adult frogs are predatory, primarily consuming insects and other invertebrates, but can eat small vertebrates. Tadpoles eat periphyton and phytoplankton that are scraped from aquatic vegetation. Tadpoles may also scavenge nutrients from dead tadpoles.

CONSERVATION AND MANAGEMENT Gopher Frogs are recognized by ADCNR as Priority 1 (species of highest conservation concern) and are protected by Alabama's Nongame Regulation (Shelton-Nix 2017).

They are also listed as vulnerable by CITES and are being considered for protection under the ESA. For these reasons, it is illegal to collect this species within Alabama without a special permit. This species is protected because it is rare throughout its geographic range, and a significant portion of that range occurs in Alabama. Viable populations appear to remain only in the Conecuh National Forest, but only eight breeding sites are known there. Unfortunately, some sinkhole breeding sites selected by Gopher Frogs as primary breeding sites are also prized areas by local fishermen who stock these ponds with sunfishes, predators that are capable of consuming all the eggs and larvae of Gopher Frogs. This loss of recruitment is known to occur even though Gopher Frog tadpoles increase hiding behavior in the presence of fish predators (Gregoire and Gunzburger 2008). Ponds used by Gopher Frogs frequently dry during drought periods and, therefore, do not harbor fishes unless assisted by humans. Gopher Frogs appear to recognize the presence of these fishes and skip reproduction when the fish predators are present. At the Conecuh National Forest, Nellie Pond was pumped dry to eliminate sunfishes, and this site was immediately reused by Gopher Frogs for reproduction, with a dramatic increase in egg masses occurring after fish removal (Jensen et al. 2003). Therefore, regulation of fish predators at known reproductive sites is vital to retaining this species. Because these frogs show high site fidelity, maintenance of open pine forests with an understory of pyrogenic grasses is essential for habitat management in areas surrounding known reproductive sites. Frequent fire is needed to maintain high-quality habitat (Roznik and Johnson 2009), but these fires should be implemented after May 1 to avoid harming frogs as they migrate to and from breeding sites (Humphries and Sisson 2012). Because the species may migrate long distances to reach breeding sites (Franz et al. 1988; Humphries and Sisson 2012) and avoid hardwood areas during these migrations (Roznik et al. 2009), buffers of at least 0.6 mi (1 km) around breeding sites are needed to maintain the open pine stands required by Gopher Frogs. Burrows of Gopher Tortoises are important refugia for Gopher Frogs outside of the breeding season, as are stump holes; maintenance of these refugia is also important for conservation of these anurans. Insecticides should not be used near Gopher Frog reproductive sites, and herbicide use should be limited to applications given under exceptional circumstances required to improve overall habitat quality. This species experiences mass mortality events that are associated with high prevalence of ranavirus (Sutton et al. 2014).

Taxonomy We follow Richter and Broughton (2005) in restricting this species to frogs of the Coastal Plain east of the Mobile drainage. It is the sister species to *L. sevosus* (Hillis and Wilcox 2005). We recognize no subspecific variation. However, three mitochondrial lineages are known (Richter et al. 2014), with Alabama specimens belonging to a northern lineage including specimens from Georgia, South Carolina, North Carolina, and the panhandle of Florida.

Adult Dusky Gopher Frog (*Lithobates sevosus*), Harrison County, MS.

Dusky Gopher Frog
Lithobates sevosus (Goin and Netting, 1940)

DESCRIPTION This is a rather large, stout-bodied frog, attaining a maximum snout–vent length of about 3.9 in (100 mm). The hind feet are extensively webbed, and the tips of the digits are not expanded. A pair of dorsolateral folds is present on the dorsum, and in calling males a pair of vocal sacs expand laterally from the throat. The dorsolateral folds may be inconspicuous because of the rough, warty skin of this species. The tympanum and eye are about the same size. The dorsal ground color is gray or light brown with large dark blotches between which are interspersed smaller dark markings of varying shape. The venter is white with numerous small spots, many of which coalesce. The groin and inner surfaces of the thighs are tinged with yellow in live specimens. Males develop enlarged cornified regions at the base of the thumb during the breeding season and have a tympanum that is larger than that of females. The tadpole is light green with diffuse dark mottling on the tail fins.

In morphology and color pattern adults and tadpoles of this species are indistinguishable from Gopher Frogs (*L. capito*). However, the relatively short hind legs and warty body make adults of these two species easily distinguishable from all other ranids in Alabama except Northern Crawfish Frogs (*L. areolatus circulosus*), which lack spots on the center of the chin and have dark dorsal spots that are outlined by light

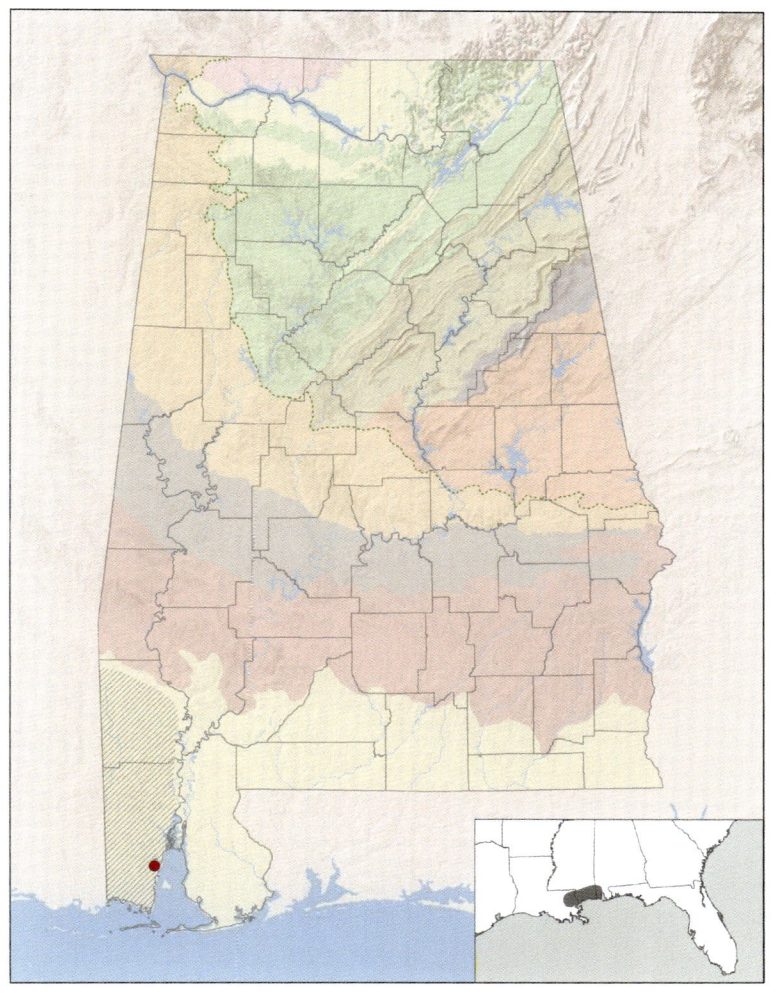

Distribution of Dusky Gopher Frog (*Lithobates sevosus*). The presumed range of the species in Alabama is indicated by hatching. Solid dots indicate localities of specimens or photographs examined by the authors or ADCNR/Natural Heritage Program occurrence records believed to be valid. Inset map depicts approximate range in the United States.

gray or tan. Tadpoles of Northern Crawfish Frogs, Gopher Frogs, and Dusky Gopher Frogs are most like those of Southern Leopard Frogs (*L. sphenocephalus*), but that species is darker green in body color and has darker mottling of the tail fin.

ALABAMA DISTRIBUTION This species is documented from the state only by a potential sight record presented in Löding (1922) for specimens found under beach debris at the mouth of the Dog River in Mobile County. However, suitable habitat is present elsewhere in Mobile County and in Choctaw and Washington Counties where we expect this species still survives. Specimens from Alabama need to be examined via DNA sequencing and electrophoresis to confirm their identity.

Habits There is little information on the habits of this shy frog. Studies in Mississippi document that adults move up to 328 yd (300 m) from the breeding site over a period of two days and spend the rest of the active season in a stump hole, root mound, or rodent burrow (Richter et al. 2001). During this active season adults may emerge nightly and remain at the burrow entrance where feeding may take place. Adults migrate to the breeding site during winter rainy periods, generally between November and January. However, not all adults migrate each year, so breeding populations are characterized by high turnover rates among years (Richter and Seigel 2002). Males arrive at the reproductive site earlier than do females and attempt to attract mates by giving a deep snoring call that is frequently produced underwater. Females deposit eggs in globular masses under water by attaching them to the stems of aquatic vegetation. Extreme variation in reproduction occurs among years because of variation in adult migration, hydroperiod, and juvenile survival (Richter et al. 2003).

Adult frogs are predatory, primarily consuming insects and other invertebrates, but are capable of eating small vertebrates. Tadpoles eat periphyton and phytoplankton that are scraped from aquatic vegetation. Tadpoles may also scavenge nutrients from dead tadpoles.

Conservation and Management In 2001, the Dusky Gopher Frog was listed as endangered under the Endangered Species Act, and this species is listed as critically endangered by CITES. Only three natural breeding sites in Mississippi are known to occur, but suitable habitat remains elsewhere in Mississippi and in Alabama. Therefore, Dusky Gopher Frogs are recognized by ADCNR as Priority 1 (species of highest conservation concern) and are protected by Alabama's Nongame Regulation (Shelton-Nix 2017). For this reason, it is illegal to collect this species within Alabama without a special permit. Suitable habitat remains in Alabama, especially in Washington and Choctaw Counties, and we expect that breeding populations eventually will be found.

A recovery plan for this species was approved in 2015 (US Fish and Wildlife Service 2015) and was modified in 2019 (US Fish and Wildlife Service 2019). These documents identify the discovery and/or creation of six metapopulations distributed across areas presumed to characterize the natural geographic range of the species as being necessary to down list these frogs to threatened status; discovery or creation of an additional four metapopulations is necessary to delist the species. Metapopulations must consist of at least two breeding ponds, and evidence of persistent reproduction at each pond over at least a ten-year period is required to indicate likely viability of each metapopulation. Additionally, each pond must be surrounded by habitat that is under management agreements that assure retention of habitat characteristics preferred by adult frogs (open pine forests managed with fire to maintain herbaceous understory plants). Though an Alabama metapopulation is not required for recovery, the presence of apparently suitable habitat in the state suggests that the state will play a role in recovery.

Research at remaining Mississippi breeding ponds will guide conservation efforts. Based primarily on studies of Glen's Pond in Mississippi, the best remaining population of Dusky Gopher Frogs, a 0.6 mi (1 km) zone of natural pine forest around each breeding site is recommended to ensure sufficient upland habitat for adults (Richter et al. 2001). The Glen's Pond population is characterized by extreme variability in size among years, which likely is associated with a high probability of eventual extinction (Richter et al. 2003). Mortality of tadpoles caused by the protistan pathogen *Dermomycoides* (US Fish and Wildlife Service 2015) and ranavirus (Sutton et al. 2014) plays an important role in population variability. Additionally, this population has reduced genetic variability, suggesting a severe bottleneck (Richter et al. 2009). To improve juvenile survival, half of recent egg masses have been raised to metamorphosis in experimental ponds. The transformed frogs are then released at the remaining reproductive site. Genetic evaluation of this population suggests that few alleles are being lost due to drift, a finding that suggests management techniques are maintaining population viability (Hinkson and Richter 2016). Sufficient egg masses and tadpoles are produced at Glen's Pond to justify recent translocation of larvae to three sites in Mississippi that are managed to improve site quality for Dusky Gopher Frogs. Additionally, habitat restoration at one pond site adjacent to Glen's Pond has yielded natural colonization of the site by reproductive adults.

If a population of Dusky Gopher Frogs exists in Alabama, it would receive protection under the Endangered Species Act and would require immediate efforts to protect its habitat. Movement of microbial pathogens among breeding sites is a concern for this species (Seigel and Dodd 2002) and use of bleach to disinfect clothes and equipment would be required to prevent pathogen transfer. Habitat surrounding any Alabama population should be managed to retain open, grass-dominated pine stands by administering frequent low-intensity growing-season fire. Burrows of rodents and stump holes are likely to be important refugia for Dusky Gopher Frogs outside of the breeding season; maintenance of these refugia is also important for conservation of these anurans.

TAXONOMY We follow Richter and Broughton (2005) in restricting this species to those gopher frogs found in the Coastal Plain west of the Mobile River. Its sister species, Gopher Frogs (*L. capito*) (Hillis and Wilcox 2005), occurs on the eastern side of the Mobile River, and this pair of species is part of the *areolatus* subgroup (Crawfish Frogs [*L. areolatus*], Gopher Frogs, Pickerel Frogs [*L. palustris*], and Dusky Gopher Frogs) of leopard frogs of the *pipiens* group (Hillis and Wilcox 2005). We consider Dusky Gopher Frogs to have no subspecific variation.

Adult Northern
Crawfish Frog,
(*Lithobates areolatus circulosus*),
Sumter County, AL.

Northern Crawfish Frog

Lithobates areolatus circulosus (Rice and Davis, 1878)

DESCRIPTION This is a rather large, stout-bodied frog, attaining a maximum snout–vent length of about 4.3 in (110 mm). The hind feet have extensive webbing between the toes, and the toes are pointed, lacking expanded toe pads. The diameter of the tympanum is similar to that of the eye, and dorsolateral ridges are present. In calling males, the vocal sacs are paired and emerge laterally from under the throat. Additionally, reproductive males possess a swollen base to each thumb. The body is rugose due to the presence of enlarged wart-like structures that may obscure the dorsolateral folds. In color, the dorsum is gray to light brown with large dark gray or purple spots that are outlined with tan or light gray. The venter is white, becoming yellowish green posteriorly. The tadpole is light green with diffuse dark mottling on the tail fins.

This species is similar in color, shape, and size to Gopher Frogs (*L. capito*) and Dusky Gopher Frogs (*L. sevosus*) but differs from those species in lacking gray spots on the chin and possessing a light border to the dark dorsal spots. Tadpoles of this species are indistinguishable from Gopher Frogs and Dusky Gopher Frogs. Tadpoles of these three are most like those of Southern Leopard Frogs (*L. sphenocephalus*), but that species is darker green in body color and has darker mottling of the tail fin.

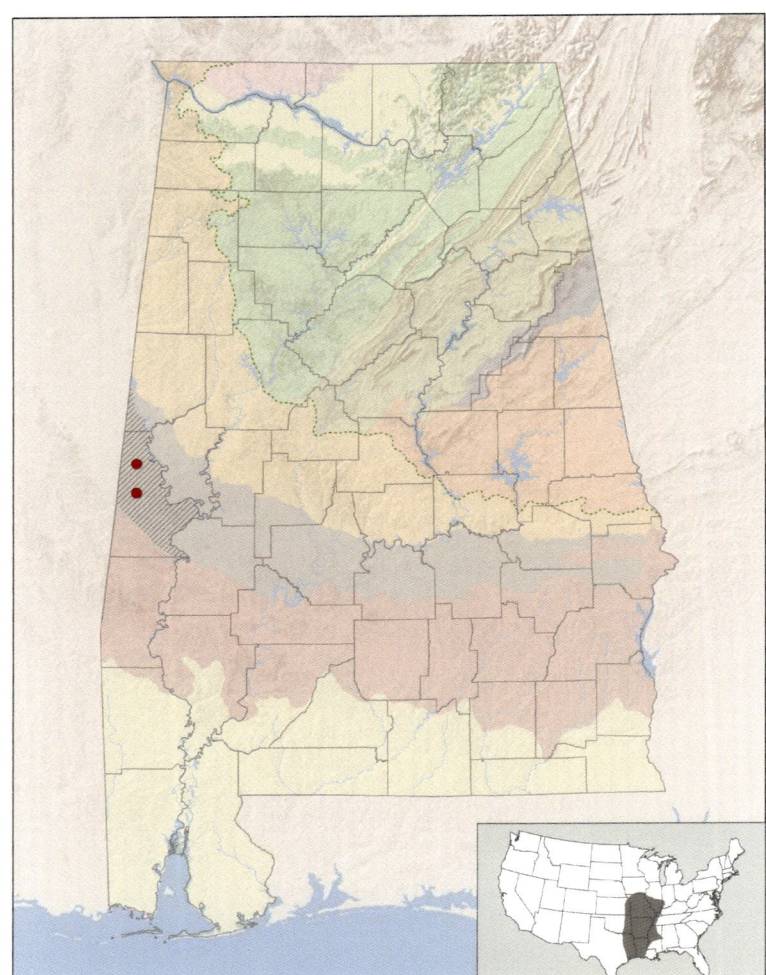

Distribution of Northern Crawfish Frog (*Lithobates areolatus circulosus*). The presumed range of the species in Alabama is indicated by hatching. Solid dots indicate localities of specimens or photographs examined by the authors or ADCNR/Natural Heritage Program occurrence records believed to be valid. Inset map depicts approximate range in the United States.

ALABAMA DISTRIBUTION Northern Crawfish Frogs were first documented from field recordings of calling males, with subsequent documentation of two populations (Holt 2015). The subspecies is known only from Black Belt soils in Sumter County, but it likely occupies extreme southwestern Pickens County as well. We chose the Tombigbee River as a likely barrier, but Greene, Hale, and Marengo Counties deserve surveys in case the species has crossed this barrier.

HABITS Northern Crawfish Frogs occupy grassland areas where clay soils create scattered pools that retain water during winter rains.

These frogs are difficult to detect during most of the year because each individual frog occupies a crayfish burrow from which it rarely moves (Heemeyer et al. 2012). This burrow retains enough water to maintain proper hydration, attracts sufficient food resources, and serves as a refuge from fires that maintain grasslands by killing encroaching hardwoods.

This subspecies breeds during the relatively warm rains of February through April. At this time, adults move up to 0.8 mi (1.2 km) to the shallow, temporary pools where breeding occurs (Heemeyer and Lannoo 2012). Males migrate first and produce a deep "waaaa" reminiscent of the call of spadefoot toads. The call attracts females to the reproductive site where they select from among calling males. A male amplexes a female by grasping her dorsally in the armpit region. The female then selects a grassy, shallow area where she deposits 4,000–7,000 eggs that are fertilized externally by the male. Tadpoles hatch in 12 days and grow to metamorphosis in 3–4 months. Adults return to the same crayfish burrow following the same route used to migrate to the breeding site (Hoffman et al. 2010).

Adult frogs eat crayfish, insects, and other frogs. Tadpoles eat periphyton and phytoplankton that are scraped from aquatic vegetation. Tadpoles may also scavenge nutrients from dead tadpoles.

CONSERVATION AND MANAGEMENT The Northern Crawfish Frog is considered to be Priority 1 (species of highest conservation concern) by the State of Alabama. Because of concern that populations of Northern Crawfish Frogs are declining elsewhere in its range (Engbrecht and Lannoo 2010), this species is considered to be vulnerable by CITES and is being considered for protection under the ESA. The primary causes of decline are loss of breeding sites due to draining of some sites and to increased hydroperiods in others, allowing predatory fishes to invade (Phillips et al. 1999). Additionally, this species is known to be affected by the chytrid fungus (*Batrachochytrium dendrobatidis*), which causes mortality in post-breeding individuals (Kinney et al. 2011). However, drought reduces the impact of this disease on Northern Crawfish Frogs (Terrell et al. 2014).

Habitat selection models for Northern Crawfish Frogs suggest that lands managed to enhance populations of this species should eliminate woody encroachment into grasslands, especially near temporary wetlands (Williams et al. 2012). This can be accomplished from prescribed fire or via mechanical removal. Because common herbicides are known to increase mortality in tadpoles, use of chemicals to control encroaching hardwoods should be avoided. Use of discs and plows should be avoided because this might be a source of direct mortality to the frogs and might alter hydrology in ways that eliminate the crayfish needed to create burrow refuges (Engbrecht et al. 2013). Fire must be used with care during early spring reproductive periods when adults migrate to reproductive sites and late summer periods when juveniles migrate from the reproductive site. Wetlands used by Northern Crawfish Frogs must be kept free of predatory fishes, which can cause reproductive failure due to consumption of eggs and tadpoles. Typically, such fish-free wetlands are created in shallow pools that dry annually. However, the soil must remain moist enough to retain crayfish.

Alabama populations of Northern Crawfish Frogs likely are restricted to private lands managed for cattle. These sites are rural and have a long history of retaining open grasslands with shallow wetlands. These features should be sufficient to retain the species in the state. However, a comprehensive survey is needed to better understand the current distribution of this species and to determine soil and topographic features, as well as the crayfish species, associated with its presence. Captive rearing of Northern Crawfish Frogs has been explored as a conservation strategy in Indiana. It has been demonstrated that survival to metamorphosis can be increased in captivity, but rates of malformed individuals can also be increased, relative to field sites (Stiles et al. 2016). Headstarting should be considered as a strategy to expand Alabama populations on to public lands managed for maintaining biodiversity.

TAXONOMY Older literature considers Gopher Frogs and Dusky Gopher Frogs to be subspecies of Crawfish Frogs due to their similar appearance, similar advertisement calls, and reliance on use of refuges created by other organisms. However, recent taxonomic works have separated these three taxa (Richter and Broughton 2005), leaving two subspecies of Crawfish Frogs, one of which occurs in Alabama. Hillis and Wilcox (2005) estimate *L. capito* + *L. sevosus* to be the sister taxa of Crawfish Frogs.

Adult River Frog (*Lithobates heckscheri*), Leon County, FL.

River Frog
Lithobates heckscheri (Wright, 1924)

DESCRIPTION River Frogs are large southeastern ranids, attaining a maximum snout–vent length of about 5.5 in (140 mm). The hind foot has extensive webbing between the toes, and the toes lack expanded toe pads. The tympanum is much larger in diameter than the eye, and the dorsum lacks dorsolateral folds. In calling males, the vocal pouch is mostly internal, and males can be distinguished from females by the enlarged tympanum (twice the diameter of the eye). In color, the dorsum is brown, occasionally with a greenish cast. The ground color of the venter is gray to grayish brown mottled with light markings. The lower lip is dark with light spots while the upper lip is trimmed with dark pigment on the lower edge and usually possesses light spots. The skin of River Frogs is noticeably bumpy, and the eye is brick red. Tadpoles of this species are distinctive in having a dark gray body, a tail with dark gray or black musculature, and a wide smoky gray or black border to the tail fins. Young tadpoles typically have a golden bar behind each eye. The tadpoles also have the unusual behavior of remaining in a tight cluster of individuals as they grow.

River Frogs are most like American Bullfrogs (*L. catesbeianus*) and Pig Frogs (*L. grylio*), but those two species have smooth skin and light bellies that are mottled with gray. River Frogs also are similar in appearance to Green Frogs (*L. clamitans*), but that species possesses

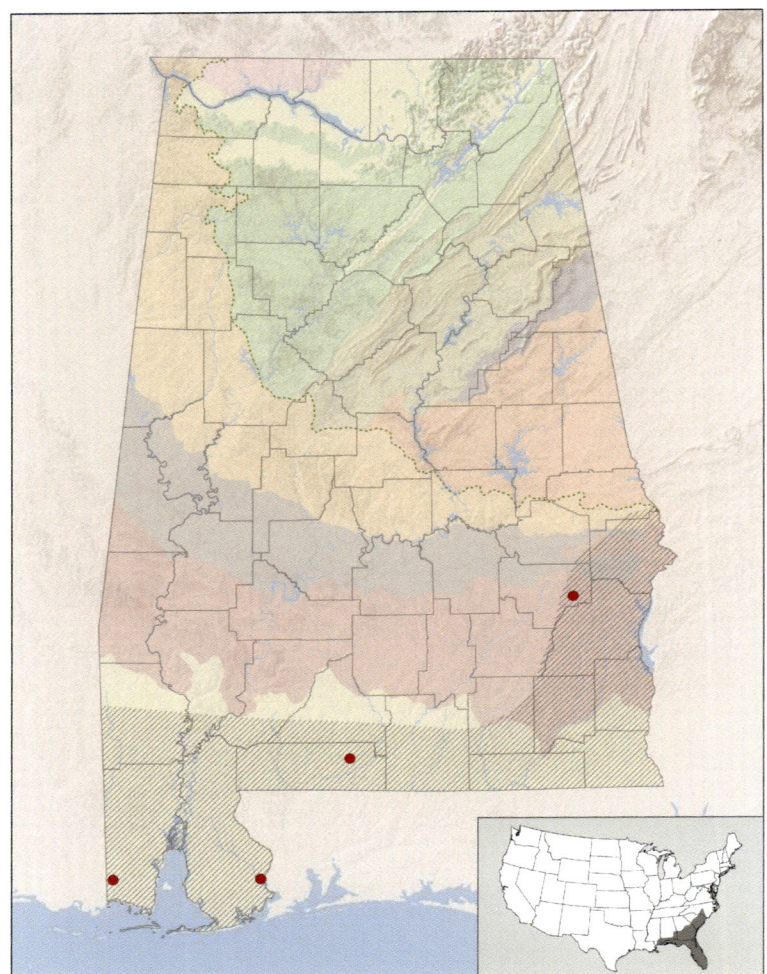

Distribution of River Frog (*Lithobates heckscheri*). The presumed range of the species in Alabama is indicated by hatching. Solid dots indicate localities of specimens or photographs examined by the authors or ADCNR/Natural Heritage Program occurrence records believed to be valid. Inset map depicts approximate range in the United States.

dorsolateral folds. The color pattern and behavior of River Frog tadpoles make them easily distinguishable from all other tadpoles.

ALABAMA DISTRIBUTION River Frogs are found along the southern tier of counties, and their range extends along the eastern boundary of the state at least to Barbour County.

HABITS The common name of this frog is misleading because it actually prefers swampy margins of smaller streams and the edges of shallow impoundments, such as beaver ponds, where titi, bay, and cypress are the dominant vegetation. River Frogs, members of a nocturnal species, are remarkably less wary than other ranids. When captured and held in hand, they usually become limp and do not struggle to free themselves. Indeed, this characteristic will help identify the species in the field.

There are no data on River Frog reproduction from Alabama. Based on data from surrounding states, the species is known to breed from April through August, in the same aquatic sites that it occupies the rest of the year. The call is variable and has been described as a snort, snarl, grunt, and guttural roll. Nothing is known about mate choice in the species, but once amplexed by males, females lay eggs as a surface film that is fertilized externally by the male. The tadpoles move about in schools, a feature shared with no other ranid in Alabama, and are notable for the exceptionally large size they attain (up to about 6.3 in [160 mm] in length), a size reached at about one year of age. During unusually cold weather this species is likely to remain in water.

Adult frogs are predatory, primarily consuming insects and other invertebrates, but are capable of eating small vertebrates. Tadpoles eat periphyton and phytoplankton that are scraped from aquatic vegetation. Tadpoles may also scavenge nutrients from dead tadpoles.

CONSERVATION AND MANAGEMENT Because they are found in few localities in the state, River Frogs are recognized by ADCNR as Priority 1 (species of highest conservation concern) and are protected by Alabama's Nongame Regulation (Shelton-Nix 2017). The species is rare both from the perspective of likely low numbers of populations in the state and the size of these populations. Surveys of remaining populations are needed to understand the status of River Frogs. Healthy populations are found in Florida and Georgia, but they are patchy in their distribution. Therefore, maintenance of populations within Alabama is likely to be an important goal of retaining this species throughout its geographic range. Draining of swamps and floodplain ponds are deleterious to River Frog populations. Mass mortality of tadpoles and juveniles in association with high prevalence of ranavirus is known (Miller et al. 2011), and chytrid fungal disease (*Batrachochytrium dendrobatidis*) has been detected in this species (Rollins et al. 2013).

TAXONOMY The River Frog is sister to American Bullfrogs (*L. catesbeianus*), making it part of the *catesbeianus* group of ranids native to eastern North America (Hillis and Wilcox 2005). This species is considered to have no subspecific variation.

Pig Frog
Lithobates grylio (Stejneger, 1901)

DESCRIPTION Pig Frogs are large ranids, attaining a maximum snout–vent length of about 6.5 in (165 mm). The hind foot has extensive webbing between the toes, with webbing extending to near the tip of the longest toe. The toes are pointed, lacking expanded toe pads. The tympanum is larger in diameter than the eye, and the dorsum lacks dorsolateral folds. When males call, the vocal pouch remains mostly internal. Additionally, adult males may be distinguished from females by their greatly enlarged tympanum (twice diameter of eye). The head typically is bright green, and the dorsum is brownish to green in color, occasionally with obscure dark markings. The chin and lips lack conspicuous white spots, and the venter has a light ground color with dark spots or mottling (occasionally unmarked). The rear of the thigh usually has a longitudinal light stripe or longitudinally oriented series of light spots. The tadpole of this species is distinctive in having a green body, dark green tail musculature with large diffuse dark spots, and smoky gray tail fins with diffuse dark spots that align in rows.

Pig Frogs are easily confused with American Bullfrogs (*L. catesbeianus*), which have a more rounded snout, lack any evidence of a light stripe or linear series of spots on the back of the thigh, and have webbing that extends only to the joint attaching the terminal phalanx to the longest hind digit. River Frogs (*L. heckscheri*) are also similar in shape and color to Pig Frogs but differ in having bumpy skin and a

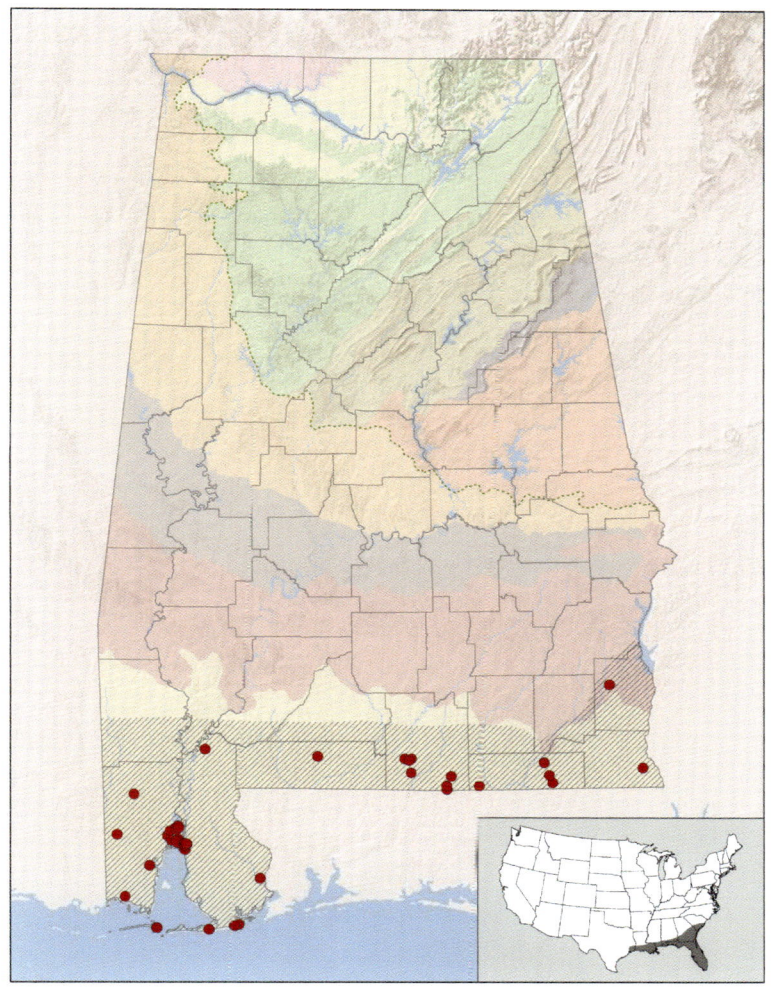

Distribution of Pig Frog (*Lithobates grylio*). The presumed range of the species in Alabama is indicated by hatching. Solid dots indicate localities of specimens or photographs examined by the authors or ADCNR/Natural Heritage Program occurrence records believed to be valid. Inset map depicts approximate range in the United States.

ventral color pattern that has a dark ground color mottled with light vermiculations. Finally, Pig Frogs are similar in appearance to Green Frogs (*L. clamitans*), but that species possesses dorsolateral folds that are not present in Pig Frogs. Tadpoles of American Bullfrogs and Pig Frogs are similar in being large and possessing a green ground color to the body. However, unlike American Bullfrogs, tadpoles of Pig Frogs have tail fins that are dark smoky gray (clear in American Bullfrogs) and lack small dorsal spots (present in American Bullfrogs, except for populations in extreme southern Houston County).

ALABAMA DISTRIBUTION Pig Frogs are found only in the southernmost tier of counties of the Lower Coastal Plain.

HABITS Permanent open bodies of water with emergent herbaceous vegetation are favored by Pig Frogs. In places where the species is abundant, such as Lake Shelby at Gulf State Park in Baldwin County, calling can be heard both day and night. The call, a loud, resonant "grunt-grunt-grunt," is mistakable only with the bellowing of calling male alligators. It is issued as the male frog floats in the water or sits among a mass of aquatic vegetation. The breeding season is March through September, when males defend territories from rival males and attract females to their territories. Thus, no migration is required for reproduction, but breeding may occur in grass and shrub areas of the wetland, avoiding hardwood aquatic zones of the same wetland used during other parts of the year. As in American Bullfrogs, male Pig Frogs are likely to avoid feeding while defending a territory and, therefore, become so weakened after several days of calling that they lose their territory to an intruder. Females likely select males based on call pitch and intensity. Once in amplexus, females lay eggs as a surface film that is up to about 19.7 in (500 mm) in diameter and is attached to vegetation. Pig Frogs do disperse across upland habitat during rainstorms, a feature that allows this species to occupy isolated wetlands not invaded by other ranid species (Delis et al. 1996). Females have higher survival rates than males, a feature that may tip adult sex ratios toward females (Woods et al. 1998). This species likely overwinters by burying itself under mud in ponds.

Adult frogs are predatory, consuming aquatic invertebrates, small fishes, and other frogs. Tadpoles eat periphyton and phytoplankton that are scraped from aquatic vegetation. Tadpoles may also scavenge nutrients from dead tadpoles.

CONSERVATION AND MANAGEMENT Pig Frogs receive no regulatory protection in Alabama. The species is abundant in south Alabama and becomes extremely abundant in peninsular Florida. It breeds in farm ponds and is abundant in Gulf State Park, the Conecuh National Forest, and Perdido Longleaf Hills Tract. Therefore, it seems secure within the state as well as throughout its geographic range. Management activities that maintain open ponds, oxbow lakes, and sluggish backwater sloughs improve habitat for this species. Urbanization does not affect occupancy patterns of this species (Guzy et al. 2012), and tadpoles are known to occupy agricultural wetlands (Babbitt and Tanner 2000).

TAXONOMY This species is the sister taxon to *L. catesbeianus* + *L. heckscheri*, making it part of the *catesbeianus* group of ranids native to eastern North America (Hillis and Wilcox 2005). The species is considered to have no subspecific variation.

Adult American Bullfrog (*Lithobates catesbeianus*), Franklin County, FL.

American Bullfrog
Lithobates catesbeianus (Shaw, 1802)

DESCRIPTION This species contains widely distributed large ranids, attaining a maximum snout–vent length of about 7.9 in (200 mm). The hind foot has extensive webbing between the toes, but it only extends to the level of the joint connecting the terminal phalanx to the longest hind digit. All toes are pointed and lack expanded toe pads, and the dorsum lacks dorsolateral folds. The diameter of the tympanum is wider than that of the eye, and the skin is smooth to slightly rough. In calling males the vocal pouch remains mostly internal. Additionally, adult males can be distinguished from females by a greatly enlarged tympanum (twice the diameter of the eye). The head of American Bullfrogs typically is green, and the dorsal ground color is greenish brown posteriorly in juveniles and is uniform brownish green in large adults. The chin and lips usually lack white spots, and there is no dark mask through the eye. The venter is off-white, usually with dark gray wormlike markings of varying shapes and intensity. The posterior portion of the thigh is spotted or mottled, and the white ground color does not form a longitudinal stripe or suggestion thereof. American Bullfrog tadpoles have a green body with small dark spots on the dorsum, and the tail fin is largely clear.

Adult American Bullfrog (*Lithobates catesbeianus*), Lauderdale County, FL.

American Bullfrogs are difficult to distinguish from Pig Frogs (*L. grylio*), but Pig Frogs have a light stripe or series of stripe-like spots on the back of the thigh, a more pointed snout, and webbing that extends beyond the joint attaching the terminal phalanx to the rest of the longest hind digit. American Bullfrogs are also like Green Frogs (*L. clamitans*), but Green Frogs possess dorsolateral folds not found in American Bullfrogs. Tadpoles of American Bullfrogs and Pig Frogs are similar in being large and the body possessing a green ground color. However, unlike American Bullfrogs, tadpoles of Pig Frogs have tail fins that are dark smoky gray with dark spots that align in rows.

ALABAMA DISTRIBUTION This species is found throughout the state. The distribution of the two mitochondrial clades (see Taxonomy section below) within the state needs to be clarified. Current evidence suggests that the Alabama River is a barrier to the eastern clade but not the western clade (Austin and Zamudio 2008).

HABITS This familiar frog is a common inhabitant of nearly all lakes and permanent ponds in Alabama, as well as many medium and large-sized streams. It also is common in marshes, cypress savannas, and cypress-gum swamps (Liner et al. 2008). Unlike most frogs in Alabama, American Bullfrogs occupy permanent wetlands that are inhabited by predatory fishes. In these wetlands, the deep, throaty call of

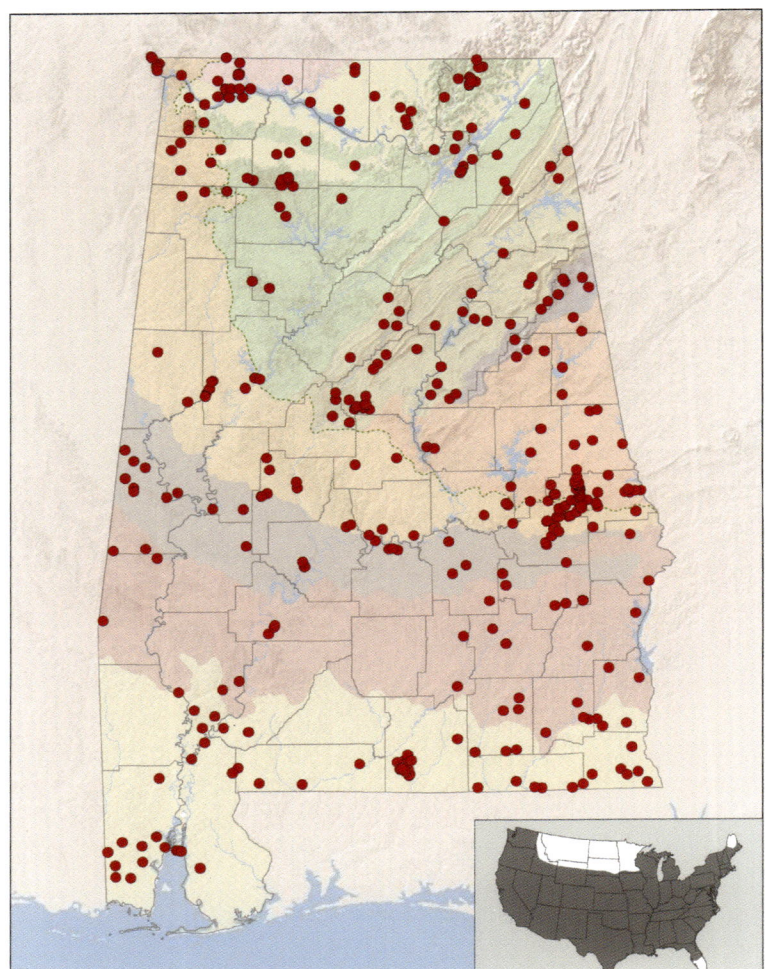

Distribution of American Bullfrog (*Lithobates cates-beianus*). Solid dots indicate localities of specimens or photographs examined by the authors or ADCNR/Natural Heritage Program occurrence records believed to be valid. Inset map depicts approximate range in the United States.

a male American Bullfrog can be heard at night from March through August. Calling during the day is not uncommon, but calling peaks from midnight to dawn, and males tend to call persistently throughout the breeding season (Bridges and Dorcas 2000). While calling, the male usually sits on a bank near water. Males defend their calling sites from intruder males and do not feed while they possess that territory. However, a territorial male eventually becomes weakened and loses his territory to an intruder. The defeated male leaves to feed and store enough energy to attempt to reestablish himself in the calling chorus (Ryan 1980). Females are known to select some males over others, but the cues used to make such choices are not known. It is known that non-calling satellite males are present in the population, and these males attempt to intercept females that are attracted to calling males (Howard 1988). Once a female selects a mate, she allows him to amplex her and deposits her eggs as a surface mass or film. The egg masses vary in diameter from 5.9–35.4 in (150–900 mm) and contain up to 20,000 eggs. In Alabama, the tadpoles attain giant size, usually by growing for over a year and then transforming during their second season. The tadpoles are unpalatable to fish (Kats et al. 1988; Adams et al. 2011), allowing the species to withstand this important source of predation. On warm, rainy fall nights small, newly transformed American Bullfrogs are often seen in abundance on highways as they disperse from reproductive sites. Current evidence suggests that these dispersing individuals are disproportionately females (Austin et al. 2003). No migration occurs in adult frogs since the areas used for breeding are the same areas occupied the rest of the year. Adult frogs overwinter by burying under mud in ponds or under leaf litter on moist soil.

American Bullfrogs are voracious feeders, capturing and swallowing almost anything that will fit in their mouths. Invertebrates constitute the bulk of the diet, but birds, snakes, turtles, mice, and other frogs, including members of its own species, may also be included. The effect of predation by adult American Bullfrogs on other frogs, especially Green Frogs, may be substantial (Werner et al. 1995). As tadpoles, American Bullfrogs consume a variety of items, including algae and aquatic vegetation that dominate the diet of most ranid tadpoles. However, American Bullfrog tadpoles also are surprisingly carnivorous in their diet (Schiesari et al. 2009).

CONSERVATION AND MANAGEMENT Because they are so common, American Bullfrogs receive no regulatory protection in Alabama regulation. The species does well in farm ponds and similar habitats created in city and county parks or golf courses; older ponds are more likely to be occupied by this species (Birx-Raybuck et al. 2009). Therefore, special management activities are not required to maintain American Bullfrogs. These frogs are harvested in some areas of the state for the meat present on the hind legs. In fact, the species has become a nuisance in some areas where it has been transported for farming purposes and escaped. It is not native to the western part of the United States but is well-established there, resulting in decreased abundance of western ranids (Kupferberg 1997), and it may have played a role in the spread of chytridiomycosis to tropical countries (Hanselmann et al. 2004). Timber harvests that create open gap areas are avoided by adults and metamorphs of this species (Strojny and Hunter 2010), and reduction of the hydroperiod of large wetlands eventually eliminates this species. Calling in males is sensitive to proximity to road noise (Cosentino et al. 2014). American Bullfrog tadpoles experience higher mortality in the presence of the fungal pathogens *Saprolegnia* and *Leptolegnia* (Ruthig 2009), but are not affected by chytrid fungal disease (*Batrachochytrium dendrobatidis*), serving instead as a carrier of this disease (Greenspan et al. 2012). Mass mortality associated with high prevalence of ranavirus is known for American Bullfrogs (Miller et al. 2011).

Taxonomy The American Bullfrog is the sister species to River Frogs (*L. heckscheri*) and is the nominate member of the *catesbeianus* group of ranids, a lineage native to eastern North America (Hillis and Wilcox 2005). No subspecies of this widespread taxon currently are recognized, but data from the mitochondrial genome document eastern and western clades that generally are separated by the Mississippi River (Austin and Zamudio 2008) but for which the western clade has invaded areas east of the Mississippi River. Specimens from both clades are found in Alabama based on a sample of the eastern clade from Lee County and of genetic material from both the eastern and western clades for specimens from Tuscaloosa County (Austin and Zamudio 2008). Based on the wide zone where both lineages co-occur, we infer that these two lineages are not on separate evolutionary trajectories and are unlikely to represent cryptic species.

Great Caribbean Landfrogs

Family Eleutherodactylidae

This large family, with 4 genera and about 215 species, is found from southern Texas south through Central America and northern South America, including the Greater and Lesser Antilles. The family is sister to the families Brachycephalidae + Craugastoridae + Strabomantidae, New World frogs of Central and South American origin that, like Eleutherodactylidae, have direct development (Heinicke et al. 2009). Members of Eleutherodactylidae are largely arboreal, and most possess expanded toe pads on all digits; these pads provide adhesion to smooth surfaces, such as leaves. Females typically deposit eggs in a terrestrial nest and then attend the nest, probably to maintain proper hydration of the eggs and to protect them from predators. Males call from leaf litter or epiphytic vegetation and use this advertisement call to attract females to their calling site. Mating rarely has been observed, but in at least one species internal fertilization occurs, a feature that may occur in all species. One species retains eggs in the oviduct of the female and gives birth to live, fully transformed offspring.

The family has three species that are native to North America, and three that have been introduced. One North American and one non-North American species have invaded Alabama.

Rainfrogs

Genus *Eleutherodactylus* (Duméril and Bibron, 1841)

We follow Hedges et al. (2008) in restricting this genus to a mono-phyletic group of small frogs that likely originated in the Greater Antilles and occupy the mainland in regions near these islands. The genus consists of 191 species, of which 3 are native to the United States and are placed in the subgenus *Syrrhophus* by some. Two species are found in Alabama, both representing recent invasions, one expanding its range from Florida and the other expanding its range from Texas. *Eleutherodactylus* is the sister genus to *Diaspora*, a radiation of small, direct-developing frogs of Central America (Heinicke et al. 2009).

KEY TO THE SPECIES OF *ELEUTHERODACTYLUS* OF ALABAMA

1a   No light spots on side of body and typically lacking light spots on lips below eye; typically lacking a dark mask from the tip of the snout through the tympanum to the attachment of the arm.

Eleutherodactylus planirostris—**Greenhouse Frog . . . page 139.**

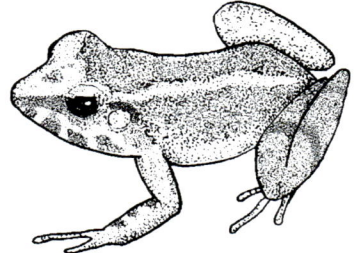

Dorsal view of Greenhouse Frog (*Eleutherodactylus planirostris*).

1b   Light spots present on side of body and on lips below eye; typically possessing a dark mask from the tip of the snout through the tympanum to the attachment of the arm.

Eleutherodactylus cystignathoides campi—**Rio Grande Chirping Frog . . . page 142.**

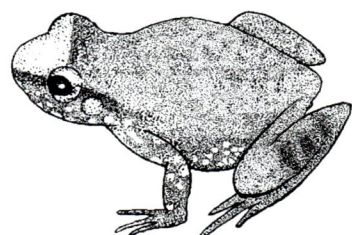

Dorsal view of Rio Grande Chirping Frog (*Eleutherodactylus cystignathoides campi*).

Adult Greenhouse Frog (*Eleutherodactylus planirostris*), South FL.

Greenhouse Frog
Eleutherodactylus planirostris (Cope, 1862)

DESCRIPTION Greenhouse Frogs are tiny anurans, attaining a maximum length of 1.4 in (36 mm). An expanded disc occurs on each digit, but these represent only a slight widening of the tip. The hind feet have no webbing. The ground color is tan or brown usually with dark mottling. Three color patterns typically are exhibited, one that is relatively uniform in color, one that possesses a pair of wide, light dorsolateral stripes, and one with a thin dark W-shaped marking across the shoulders. Most individuals lack a distinct dark mask extending from the tip of the snout to the tympanum and lack white lateral spotting. The groin region and back of the thigh are orange, and the venter is uniform white.

Greenhouse Frogs are most similar in appearance to Rio Grande Chirping Frog (*E. cystignathoides*), but that species has a distinct dark mask extending from the tip of the snout to the tympanum and white lateral spotting. Greenhouse Frogs also are similar in appearance to the Cricket Frogs (*Acris gryllus* and *A. crepitans*), but Cricket Frogs possess webbing on the hind feet and do not possess expanded discs on the digits.

ALABAMA DISTRIBUTION This species was first recorded in Alabama by Carey (1982) from specimens discovered at Marietta Johnson School of Organic Education in Fairhope, Baldwin County. The species has

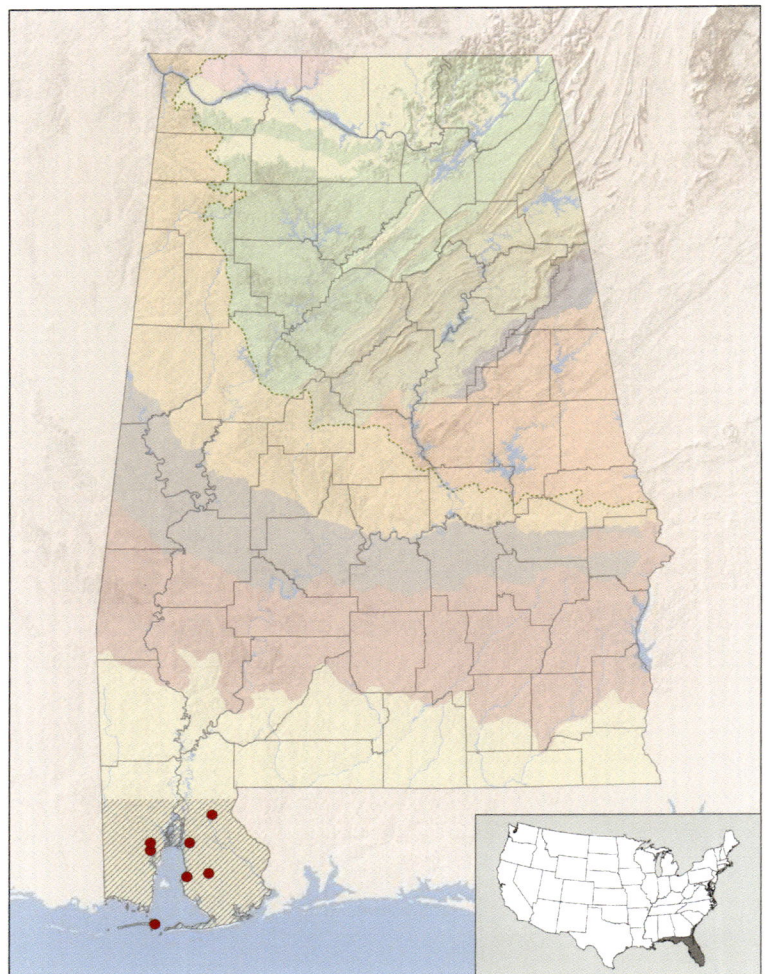

Distribution of Greenhouse Frog (*Eleutherodactylus planirostris*). The presumed range of the species in Alabama is indicated by hatching. Solid dots indicate localities of specimens or photographs examined by the authors or ADCNR/Natural Heritage Program occurrence records believed to be valid. Inset map depicts approximate range in the United States.

since been recorded from several headwater streams throughout Baldwin County (Alix et al. 2014b), suggesting that these frogs are spreading north and westward from established populations (Meshaka 2011).

HABITS Greenhouse Frogs are native to Cuba and the Bahamas, where they inhabit moist evergreen forests, either occupying epiphytes or thick, moist leaf litter. Whether the species is native to the Florida Keys or was transported there by humans is debated (Meshaka 2011). Regardless, humans appear to have assisted introduction northward from South Florida, where Greenhouse Frogs thrived first in large nurseries, but later in urban gardens and finally native habitats. They have been transported widely by hitching rides in potted plants sold in the garden industry, and this is how the species is believed to have been transported to North Florida (Meshaka 2011) and eventually Alabama. Males vocalize, producing a short single or double chirp (typically three in Rio Grande Chirping Frogs) or a short trilled chirp that sounds like an insect (more prolonged in Rio Grande Chirping Frogs). Females deposit 2–26 eggs under leaf litter between May and September and do not attend the nest (Schwartz and Henderson 1991). Based on populations of these frogs in Florida, frogs in Alabama are expected to be active except for the coldest months (January–March), to have peak recruitment from September to December, and to have a calling period of five months (May–September; Meshaka and Layne 2005). These frogs eat ants, beetles, spiders, mites, and collembolans.

CONSERVATION AND MANAGEMENT In Florida, this species has invaded native habitats, where it is thriving. It does well in open Longleaf Pine (*Pinus palustris*) savannas where it has been observed to use the moist soil in Gopher Tortoise (*Gopherus polyphemus*) burrows to avoid desiccation (Meshaka 2011). For this reason, it might be expected to expand across the southern tier of counties of Alabama. However, there is no evidence that this species competes with native frogs, and it produces no known noxious substances that would appear to cause problems for native predators. Therefore, there is no apparent need to prevent its spread within the state. Current records suggest that the spread of this species in Alabama is still largely associated with urbanized areas (Alix et al. 2014b).

TAXONOMY Four subspecies are recognized (Schwartz and Henderson 1991) but recent accounts have avoided use of them.

Adult Rio Grande Chirping Frog (*Eleutherodactylus cystignathoides campi*), South FL.

Rio Grande Chirping Frog

Eleutherodactylus cystignathoides campi (Stejneger, 1915)

DESCRIPTION The Rio Grande Chirping Frog is a tiny anuran, attaining a maximum length of 1 in (26 mm). An expanded disc occurs on each digit, but these represent only a slight widening of the tip. The hind feet have no webbing. The ground color is tan or brown with dark mottling. Most individuals have a distinct dark eye mask extending from the tip of the snout, through the eye, to the tympanum. Additionally, most individuals have white spots on the lips below the eye and along the lateral surface of the body.

Rio Grande Chirping Frogs are most like Greenhouse Frogs (*E. planirostris*), but that species has no distinct dark eye mask extending from the tip of the snout to the tympanum. Additionally, most individual Greenhouse Frogs lack white spots on the lips, below the eye, and along the lateral surface of the body. Rio Grande Chirping Frogs also are similar in appearance to the Cricket Frogs (*A. gryllus* and *A. crepitans*), but Cricket Frogs possess webbing on the hind feet and do not possess expanded discs on the digits.

ALABAMA DISTRIBUTION This species was first recorded from Alabama in 2014 from specimens discovered in trash piles west of Mobile (McConnell et al. 2015).

HABITS Rio Grande Chirping Frogs are native to the Gulf Coast of extreme southern Texas and northern Mexico. Most individuals are

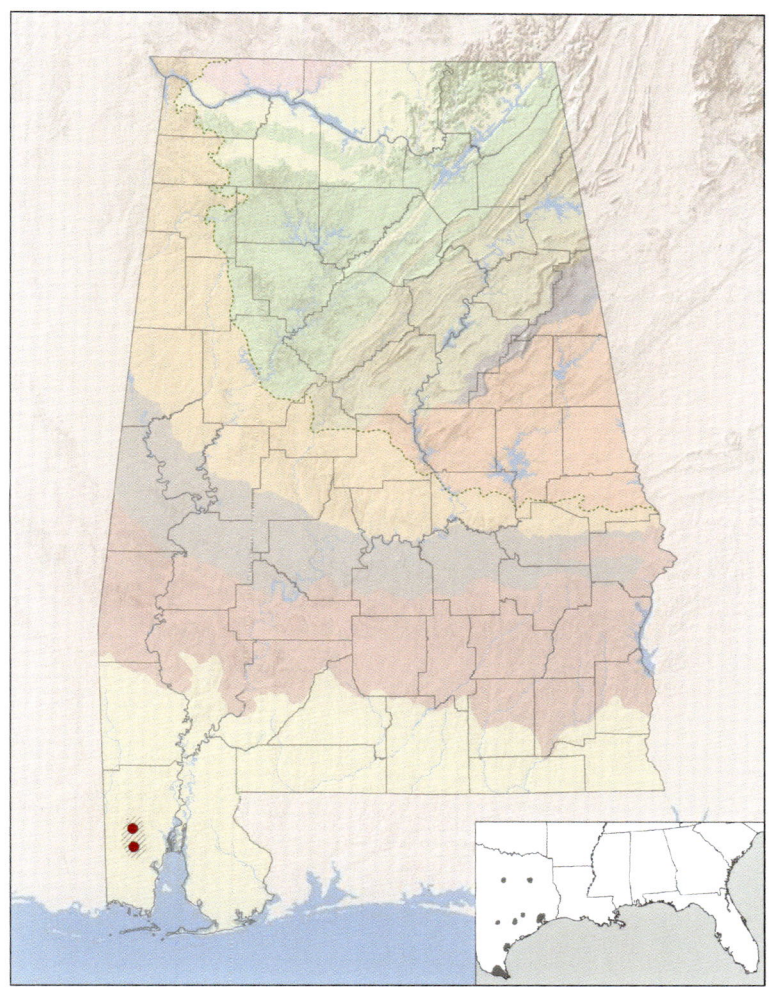

Distribution of Rio
Grande Chirping
Frog (*Eleuthero-
dactylus cystigna-
thoides campi*). The
presumed range
of the species in
Alabama is indi-
cated by hatching.
Solid dots indicate
localities of spec-
imens or photo-
graphs examined
by the authors or
ADCNR/Natural
Heritage Program
occurrence records
believed to be
valid. Inset map
depicts approxi-
mate range in the
United States.

found on the ground where they prefer areas with moist leaf litter, but individuals have been observed perched on vegetation up to 5 ft (1.5 m) above ground (Hayes-Odum 1990). These frogs are also found along riparian zones. They do well in areas disturbed by humans, especially where trash piles occur on moist soil. Males call on warm rainy days from April through May, when breeding occurs. The call of males is a series of three short chirps given in rapid succession (typically single or double in Greenhouse Frogs) or a more prolonged trill with a rising inflection (shorter and without a rising inflection in

Greenhouse Frogs). Females produce clutches of 5–13 large eggs that are visible through the skin of the abdomen and, once laid, are not attended by the female. The eggs incubate for 14–16 days, hatching as tiny froglets. Unlike our native frogs, Rio Grande Chirping Frogs do not form breeding aggregations, instead establishing territories that serve as centers of all activities, including reproduction (Hayes-Odum 1990). The diet of this species is poorly known, but likely includes ants, mites, and collembolans.

CONSERVATION AND MANAGEMENT This species is rapidly invading low-lying areas throughout the Gulf Coast. The species has vastly expanded its range in Texas and is now known from Louisiana, Mississippi, Alabama, and Florida. It likely invades new areas as a stowaway in potted plants when they are delivered to large nurseries. The species does particularly well in gardens and other areas modified by humans. However, there is no evidence that this species competes with native frogs, and it produces no known noxious substances that would appear to cause problems for native predators. Therefore, there is no apparent need to prevent its spread within the state.

TAXONOMY This species is native to North America, belonging to the subgenus *Syrrhophus*, a lineage containing other North American endemics. Some authors elevate this subgenus to the genus status. Two subspecies are recognized of which one is found in Alabama.

Treefrogs

Family Hylidae

The family Hylidae includes the treefrogs, a diverse radiation of anurans that occurs in North and South America, northern Africa, Australia, New Guinea, and portions of Europe and Asia. We follow Frost et al. (2006) in restricting this family to a monophyletic lineage that includes about 46 genera and more than 900 species. Hylids are the sister family to eight leptodactyliform families that have a cosmopolitan distribution (Frost et al. 2006); these frogs are part of a diverse successful radiation. All members of the family have toe pads that attach to the digits via an intercalary cartilage—an extra segment to each toe that improves adhesion to smooth surfaces and that allows these frogs to invade arboreal habitats. Members of the family typically lay aquatic eggs that hatch into an aquatic tadpole that then transforms into an adult body form. Aquatic sites may include water that accumulates in axils of leaves or depressions in soil created by males. Three genera with 15 species are found in Alabama.

1a Webbing present on hind feet but extending no farther than base of terminal toe bone; tips of fingers without expanded discs; rear of thigh with one or more dark longitudinal stripes or suggestion thereof; front of snout with light vertical lines; right and left segments of second upper labial tooth rows of tadpole widely separated; tadpoles with dorsal eyes, laterally compressed body, and tail often with black tip.

Genus *Acris*—Cricket Frogs . . . page 148.

From left to right:

Ventral view of hind foot of *Acris* showing webbing extending no farther than base of terminal bone of fourth toe.

Ventral view of finger of *Acris* showing tips of digits lacking expanded pads.

Clockwise from top left:

Posterior view of thigh of *Acris*.

Ventral view of *Acris* tadpole mouth parts showing widely separated right and left segments of second upper labial tooth row.

Lateral view of *Acris* tadpole.

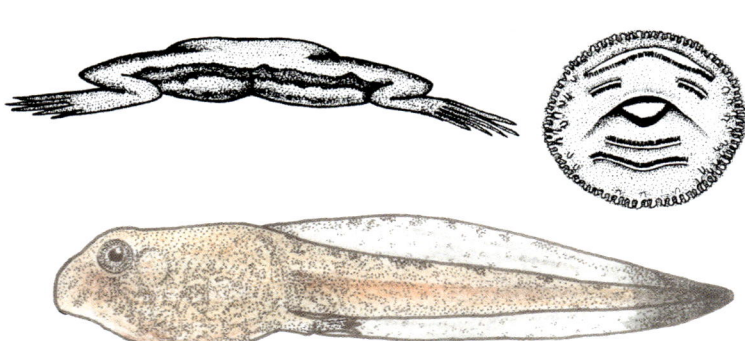

1b Webbing of hind foot either absent or extending well over half way along fourth toe (see illustrations with couplet 2); tips of fingers with expanded discs (sometimes barely wider than width of finger; see illustrations with couplet 2); rear of thigh without longitudinal stripes; front of snout without light vertical lines; right and left segments of second upper labial tooth row of tadpoles narrowly separated; tadpoles with lateral eyes, globose body, and tail lacking black tip (see illustrations with couplet 2); **go to 2.**

Ventral view of Pine Woods Treefrog (*Dryophytes femoralis*) tadpole mouth parts showing narrowly separated right and left segments of second upper labial tooth row.

2a No webbing between hind toes; tips of digits moderately expanded (to less than one-half the diameter of the tympanum), except for Spring Peeper (*Pseudacris crucifer*), which has widely expanded tips of digits; tadpoles either with dark dorsal or lateral stripe along tail musculature associated with typical tail width (dorsal fin not extending to top of head) or with noticeably wide tail (dorsal fin extending to top of head) and lacking dark central tail spot.

Clockwise from top left:

Ventral view of hind foot of *Pseudacris* showing no or minimal webbing.

Ventral view of finger of *Pseudacris* showing moderately expanded toe pad.

Lateral view of *Pseudacris* tadpole showing dark lateral stripe along tail musculature of most species.

Genus *Pseudacris*—Chorus Frogs . . . page 158.

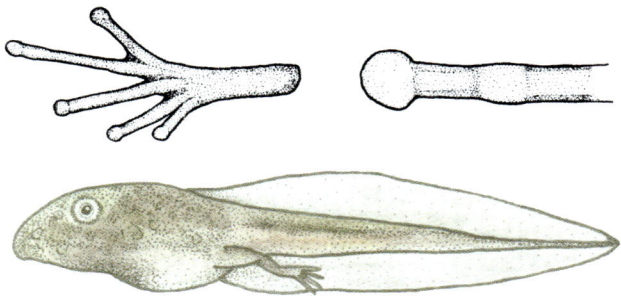

2b Webbing between hind toes to about halfway along fourth toe; tips of digits greatly expanded (to more than one-half diameter of tympanum); tadpole either lacking dark dorsal or lateral stripe along tail musculature associated with a typical tail width (dorsal fin not extending to top of head), or with a noticeably wide tail (dorsal fin extending to top of head) and with a dark central tail spot or red fin.

Clockwise from top left:

Ventral view of hind foot of *Dryophytes* showing webbing beyond base of terminal toe bone.

Ventral view of finger of *Dryophytes* showing wide expansion of toe pad.

Ventral view of *Dryophytes* tadpole lacking dark lateral stripe.

Genus *Dryophytes*—Holarctic Treefrogs . . . page 183.

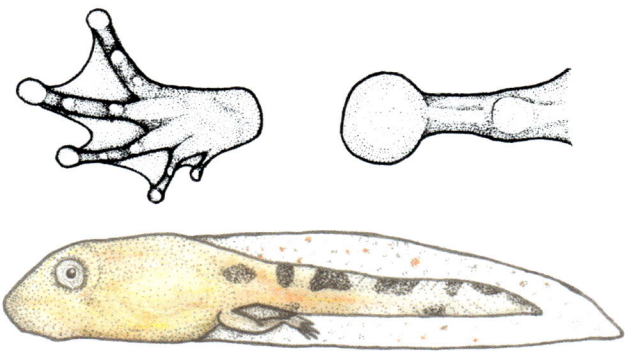

Cricket Frogs

Genus *Acris* (Duméril and Bibron, 1841)

This genus contains the Cricket Frogs of North America. These are small, largely terrestrial frogs found in forest leaf litter and on grassy aquatic margins. The common name is derived from the fact that, like crickets, these animals leap great distances and produce sounds that have similar qualities to those of some insects. Toe pads in this genus have become reduced so that the tips of the digits are not expanded relative to the rest of the digit, an adaptation to the largely terrestrial habits of these frogs. The intercalary cartilage is still present, a feature that suggests expanded toe pads were present in the ancestor of these frogs. Three species are found in the genus, two of which are found in Alabama. These frogs are the sister genus to *Pseudacris* and, along with that genus, represent a lineage of North American origin that is either sister to a rich clade of frogs of Middle American origin, which includes North American forms in the genus *Dryophytes* (Wiens et al. 2006), or is sister to the Middle American genera *Plectrohyla* and *Exerodonta* (Frost et al. 2006). Regardless, these frogs are widespread and, therefore, are a successful group.

KEY TO THE SPECIES OF *ACRIS* OF ALABAMA

1a   Body stout; snout rounded; thigh stripe usually with ragged edge or obscure; webbing on fourth toe extending at least to base of second phalanx; tadpole with light throat and mottled or reticulate tail musculature.

Acris crepitans—**Eastern Cricket Frog** . . . **page 150.**

Clockwise from top left:

Rear of thigh of Eastern Cricket Frog (*Acris crepitans*).

Ventral view of hind foot of Eastern Cricket Frog (*Acris crepitans*).

Lateral view of Eastern Cricket Frog (*Acris crepitans*) tadpole.

1b Body slender; snout pointed; thigh stripe(s) usually smooth-edged and prominent; webbing on fourth toe restricted to basal phalanx; tadpole with dark throat and finely flecked tail musculature.

Acris gryllus—**Southern Cricket Frog . . . page 154.**

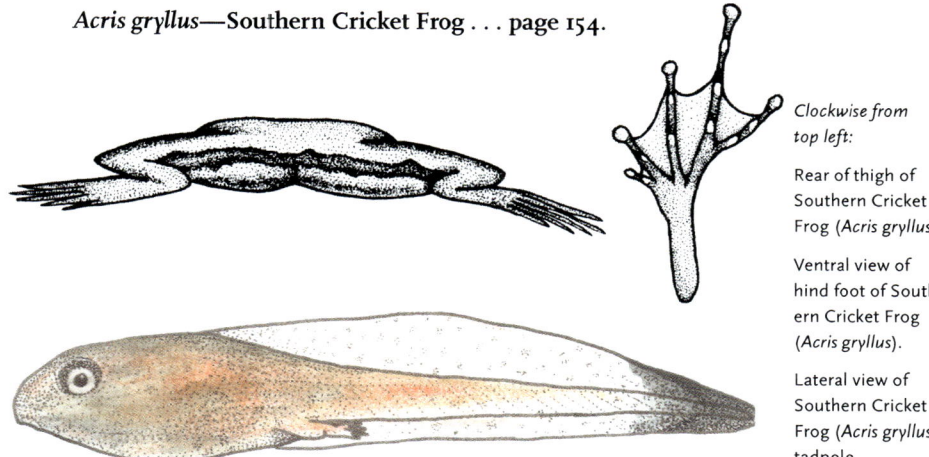

Clockwise from top left:

Rear of thigh of Southern Cricket Frog (*Acris gryllus*).

Ventral view of hind foot of Southern Cricket Frog (*Acris gryllus*).

Lateral view of Southern Cricket Frog (*Acris gryllus*) tadpole.

Eastern Cricket Frog
Acris crepitans (Baird, 1854)

DESCRIPTION Eastern Cricket Frogs are small anurans, attaining a maximum snout–vent length of about 1.4 in (35 mm). The snout is rounded, and the tips of the digits are slightly rounded but not expanded as in *Pseudacris* and *Dryophytes*. The webbing on the longest (fourth) toe extends at least to the base of the second phalanx and may extend to halfway along the second phalanx. The dorsum is warty with a ground color of gray to dark brown. The head has a dark triangular mark between the eyes, and there are variable dark spots or blotches on the rest of the dorsum, especially along the sides where a bold dark blotch is present that is surrounded by white. In some individuals there is a median green stripe along the back. The front of the snout is marked with narrow vertical light lines, and the rear of the thigh usually has a longitudinal light stripe with a scalloped border. A pair of prominent light warts usually is present on each side of the vent. Tadpoles of Eastern Cricket Frogs are medium in size, have a golden-brown cast with weak dark mottling on the tail musculature and tail fin, and in areas with dragonfly larvae predators develop a black tail tip.

Eastern Cricket Frogs are quite similar in appearance to the more southerly distributed Southern Cricket Frog (*A. gryllus*). The two can be distinguished by the narrow snout, straight boundary of the thigh stripe, less extensive webbing on the fourth toe, and lack of enlarged light warts on either side of the vent in Southern Cricket Frogs. Tadpoles of Southern Cricket Frogs also are quite similar to those of Eastern Cricket Frogs but lack weak dark mottling on the tail musculature

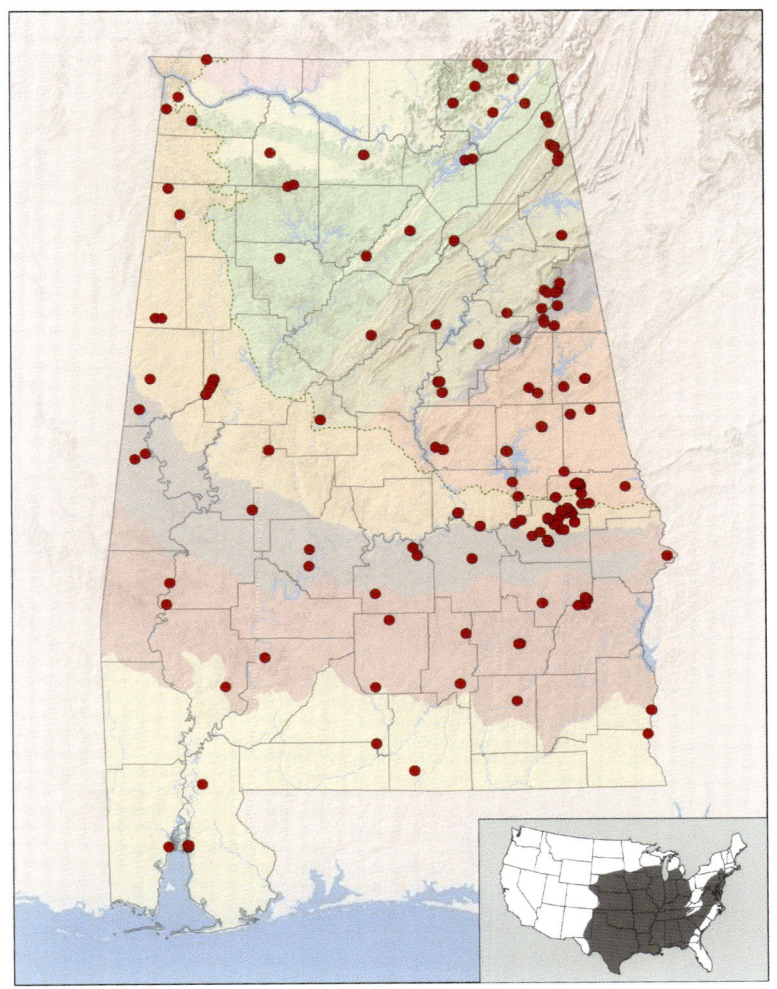

Distribution of Eastern Cricket Frog (*Acris crepitans*). Solid dots indicate localities of specimens or photographs examined by the authors or ADCNR/Natural Heritage Program occurrence records believed to be valid. Inset map depicts approximate range in the United States.

and fin. Additionally, the call of males of Southern Cricket Frogs is less raspy with more staccato than that of Eastern Cricket Frogs.

ALABAMA DISTRIBUTION Eastern Cricket Frogs, or their influence in apparently hybrid populations, occur essentially statewide. The species is abundant in most areas above the Fall Line and is locally common in portions of the Upper Coastal Plain. However, in south Alabama the species is much less abundant and is found mostly along the floodplains of major rivers. It appears to be absent from large areas of extreme southeastern and southwestern Alabama.

HABITS The Eastern Cricket Frog occurs in a wide variety of aquatic habitats, including streams, ponds, lakes, and floodplain pools. Unlike Southern Cricket Frogs, Eastern Cricket Frogs shun places that dry periodically, and they are not usually common at sites where the substrate is covered by dense, low vegetation. Mudflats and sparsely vegetated or barren shorelines are preferred microhabitats, especially those that have gently sloping banks leading to shallow water with submerged vegetation (Beasley et al. 2005). Such sites are used both as breeding sites and as areas used during the remainder of a season of activity. Eastern Cricket Frogs are decidedly less wary and less active than Southern Cricket Frogs. As a result, Eastern Cricket Frogs are more easily captured. Breeding occurs from March through July or August. The advertisement call of male Eastern Cricket Frogs consists of a series of raspy "rik, rik, . . . rik" notes. Males call in choruses in which their behavioral response to intruder males depends on local density. When local density is low, males defend calling sites against intruders by wrestling with them; in high density choruses males tolerate intruders. Males are more likely to attack intruders late in the mating season than early in it (Burmeister et al. 1999). Males indicate aggressive intent to intruder males by lowering the dominant frequency of their voice and increasing call complexity; females are more attracted to males giving such calls than calls of isolated males (Kime et al. 2004). Dominant frequency carries information about male fighting ability (large males with deep voices typically win fights) and call complexity carries information about aggressive intent (complex calls indicate willingness to fight; Burmeister et al. 2002). Calling males are site specific, returning to it when displaced (Perrill and Shepard 1989). Individuals captured on land and released in water use Y-axis orientation to swim in the direction of the homeward shore (Ferguson et al. 1967). Females are attracted to low dominant frequencies, increased number of pulses per call, and number of pulse groups per call (Ryan et al. 1995). After selecting a mate, females are amplexed by the male and select a nest site where up to 400 eggs are deposited in a series of small clusters that are fertilized externally by the male. For the remainder of the year, adult frogs may be found active in any months, but likely burrow into grass clumps or leaf litter during unusually cold periods.

Flying and ground arthropods are the primary diet items of adults, and the diet items are so diverse that limited selectivity from among available prey is exhibited in the diet (Labanick 1976). Tadpoles consume algae and scrape aquatic vegetation. Tadpoles typically have black tail tips, but those in lakes and streams may lack this feature (Caldwell 1982). These larvae are marginally unpalatable to predators (Adams et al. 2011).

CONSERVATION AND MANAGEMENT Eastern Cricket Frogs are abundant in the northern half of the state and because of this receive no regulatory protection in Alabama. The species can be found around farm, golf course, and park ponds if wooded vegetation is present. Thus, the species survives well in areas altered by humans. No special management tools appear to be required to maintain these frogs. However, populations of Blanchard's Cricket Frogs (*A. blanchardi*) and Eastern Cricket Frogs have declined in northern states where increased numbers of abnormal individuals have been observed (McCallum and Trauth 2003) and where polychlorinated biphenyls (PCBs) are thought to have caused endocrine disruption (Reeder et al. 2005). Given these population trends in other states, this common frog deserves to be monitored in Alabama. Chytrid fungal disease (*Batrachochytrium dendrobatidis*) has been detected in this species (Saenz et al. 2010), but mass mortality events have not been associated with it. Similarly, ranavirus has been detected in Eastern Cricket Frogs (Miller et al. 2011) but without association with mass mortality events.

TAXONOMY We follow Gamble et al. (2008) in recognizing Blanchard's Cricket Frog as a separate species from Eastern Cricket Frogs and in including Coastal Cricket Frog (*A. c. paludicola*) within Blanchard's Cricket Frogs. However, differentiating Blanchard's Cricket Frog from Eastern Cricket Frogs based on external features is difficult (McCallum and Trauth 2006). Furthermore, evidence from the nuclear and mitochondrial genomes suggest an eastern and a western lineage within Eastern Cricket Frogs that does not correspond to any morphological features but appears to have biogeographic significance because the western form is found north of the Ohio River (Gamble et al. 2008). The eastern lineage is documented for the state of Alabama by Gamble et al. (2008). The western clade is not known from the state but might be expected in the northern tier of counties, north of the Tennessee River.

Adult Southern Cricket Frog (*Acris gryllus*), Okaloosa County, FL.

Southern Cricket Frog
Acris gryllus (Le Conte, 1825)

DESCRIPTION The Southern Cricket Frog is a small, extremely variable anuran, attaining a maximum snout–vent length of about 1.2 in (30 mm) and with a sharply pointed snout. The tips of the digits are slightly rounded, but not expanded as in *Pseudacris* and *Dryophytes*. The longest (fourth) toe on the hind foot usually has at least two phalanges free of webbing. The dorsum is slightly rugose but not warty. The dorsum is extremely variable with a ground color of gray, brown, or black. A yellow, green, or red middorsal stripe usually is present, sometimes bifurcating anteriorly, with a fork extending to each eye. The head has a dark triangular mark between the eyes, and there are variable dark spots or blotches on the rest of the dorsum, especially along the sides where a bold dark blotch is present that is surrounded by white. The front of the snout is marked with narrow vertical light lines. On the rear of the thigh a longitudinal light stripe that has a straight border usually is present with a less distinct dark stripe below it; no pair of prominent light warts is found on each side of the vent. Tadpoles of Southern Cricket Frogs are medium in size, have a golden-brown cast, and lack any dark mottling on the tail musculature and tail fin; in areas with dragonfly larvae as predators, these tadpoles develop a black tail tip.

Positive identification of the two species of Cricket Frogs in Alabama is often difficult. The Southern Cricket Frog can be distinguished from Eastern Cricket Frogs (*A. crepitans*) by the round snout,

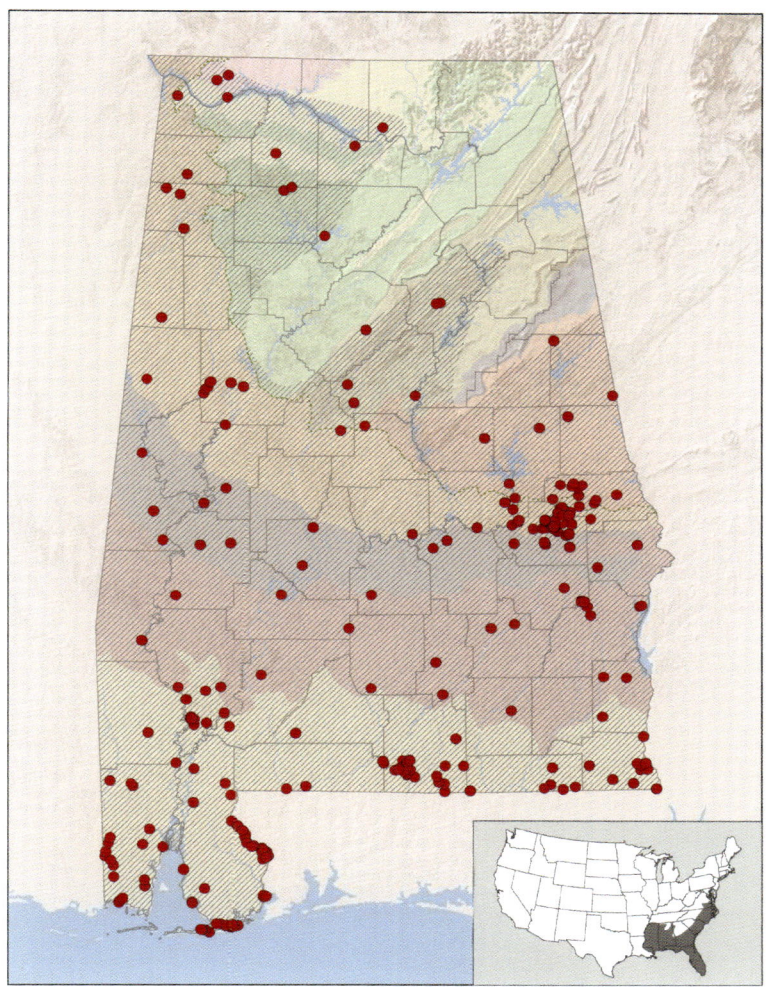

Distribution of Southern Cricket Frog (*Acris gryllus*). The presumed range of the species in Alabama is indicated by hatching. Solid dots indicate localities of specimens or photographs examined by the authors or ADCNR/Natural Heritage Program occurrence records believed to be valid. Inset map depicts approximate range in the United States.

more extensive webbing on the fourth hind toe, dark stripe with scalloped border on the back of the thigh, and presence of a light wart on each side of the vent in Eastern Cricket Frogs. In the tadpole stage, Eastern Cricket Frogs have diffuse dark mottling to the tail musculature and fin that is not present in Southern Cricket Frogs.

ALABAMA DISTRIBUTION Southern Cricket Frogs are common below the Fall Line and penetrate extensively into the Piedmont as well as the Ridge and Valley formations. The species is also present in the mountains north of the Tennessee River in Jackson and Madison Counties.

Populations of Southern Cricket Frogs in portions of the Coastal Plain above the Red Hills region may hybridize with Eastern Cricket Frogs.

HABITS The Southern Cricket Frog may be found virtually year-round in Alabama. It occurs around most kinds of permanently aquatic habitats as well as around nearby temporary accumulations of water (Liner et al. 2008) and is present at these sites for both breeding purposes and for all activities outside the breeding season. This species, in contrast to Eastern Cricket Frogs, thrives in densely vegetated places with little or no exposed substrate, such as those provided by the edges of shallow sinkhole ponds and grassy bogs and marshes. They occasionally are encountered in forested uplands surrounding such areas. In regions above the Fall Line, the Southern Cricket Frog will more likely be found in areas where the soil is relatively sandy than in heavy-soil areas. The Southern Cricket Frog is an excellent jumper and is considerably more active and elusive than the Eastern Cricket Frog. It avoids open water when attempting to escape, and if forced to jump into open water, it may jump immediately back onto land or into concealing vegetation. During winter this species likely seeks cover at the base of grass clumps or under leaf litter to avoid unusually cold periods.

Males call from mats of floating or emergent vegetation in the water or from protected places along the shore. The call is a series of "gik-gik . . . gik" notes that resemble the sound of two rocks struck together. Although choruses of males peak in intensity at 2:00 a.m., males may also call throughout all hours of the day and night and call relatively persistently throughout the breeding season (Bridges and Dorcas 2000). Call detection increases on rainy days (Steen et al. 2013). A total of up to 250 eggs may be produced by a female during a reproductive season. The eggs are deposited singly or in small clusters that are attached to stems of aquatic vegetation or sink to the bottom. Nest sites appear to be adapted to minimize contact with aquatic predators (Kats et al. 1988).

As in Eastern Cricket Frogs, flying and ground arthropods likely are the primary diet items of adults. Tadpoles consume algae and scrape aquatic vegetation. Tadpoles typically have black tail tips, and larvae are marginally unpalatable to predators (Adams et al. 2011).

CONSERVATION AND MANAGEMENT Southern Cricket Frogs are abundant in the southern half of the state and because of this receive no regulatory protection in Alabama. The species can be found around farm, golf course, and park ponds. Thus, these frogs survive well in areas altered by humans. However, occupancy of this species is decreased in urbanized areas, and its presence indicates healthy wetlands (Guzy et al. 2012). Occupancy is also reduced for wetlands associated with agriculture (Alix et al. 2014a). Management that retains open, herbaceous areas around wetlands will maintain this species. Chytrid fungal disease (*Batrachochytrium dendrobatidis*) has been detected in this species (Rizkalla 2010), but not in association with mass mortality. Populations from the northern edge of the geographic range (North Carolina and Virginia) have been lost over time (Micancin and Mette 2009).

TAXONOMY This species is sister to *A. crepitans* + *A. blanchardi*. Two subspecies of Southern Cricket Frogs traditionally are recognized. One of these, the Florida Cricket Frog (*A. g. dorsalis*), is predominantly Floridian, while the Coastal Plain Cricket Frog (*A. g. gryllus*) occurs throughout the rest of the range of the species. However, examination of the nuclear and mitochondrial genomes document separation of an eastern clade from a western clade, with the Mobile drainages serving as an apparent dispersal barrier. The eastern clade is documented from Alabama, and the western clade is suspected to occur in Mobile and Washington Counties (Gamble et al. 2008). A single specimen from Sumter County, Florida, provided limited support for Florida Cricket Frog's presence in Alabama, but the occurrence of this lineage in Alabama is doubtful because specimens from Walton and Santa Rosa Counties in the Florida Panhandle cluster with those from Coffee and Covington Counties of Alabama (Gamble et al. 2008). From these recent data we conclude that the Florida Cricket Frog remains a viable taxonomic category but that its influence is not found in Alabama. Instead, we assign all specimens from Alabama to *A. g. gryllus*. Because we find no morphological character that separates eastern from western clades of this subspecies, we give them no formal taxonomic status. However, we note that the type specimen of *Acris gryllus* var. *bufonia* is from Louisiana (Frost 2017), and this name might be assigned to the western clade of Gamble et al. (2008), with *A. g. gryllus* being restricted to the eastern clade. We retain the name Coastal Plain Cricket Frog from Crother et al. (2012).

Chorus Frogs
Genus *Pseudacris* (Fitzinger, 1843)

This genus includes 18 species that occur throughout much of the United States, especially east of the Rocky Mountains, and over a fairly large portion of Canada. It is sister to the genus *Acris*, creating a lineage of hylids that is likely to be of North American origin (Frost et al. 2006, Wiens et al. 2006). Members of *Pseudacris* all have reduced toe pads, a feature that generally distinguishes these frogs from the genus *Dryophytes*. Additionally, *Pseudacris* are all relatively small, breed during winter, and typically are gray with irregular dark stripes or blotches. Four groups are recognized within this genus. These include a western clade of species from the Pacific coast of North America, a fat-frogs clade of relatively round frogs of eastern North America, a *crucifer* clade of eastern frogs with slender bodies and high-pitched calls, and a trilling clade of eastern frogs with slender bodies and pulsed, grating calls (Moriarty and Cannatella 2004). Six species are native to Alabama, including members of all groups but the western clade.

Key to the Species of *Pseudacris* of Alabama

From left to right:

Dorsal view of Little Grass Frog (*Pseudacris ocularis*).

Lateral view of Little Grass Frog (*Pseudacris ocularis*) tadpole.

1a Adult size less than 0.7 in (18 mm) from snout to vent; a dark lateral stripe passing through the eye and ending on the side of the body; tadpole with dark stripe on tail musculature, bordered below by light stripe that extends forward to eye.

  Pseudacris ocularis—**Little Grass Frog . . . page 161.**

1b Adult size over 0.8 in (20 mm); markings various but lacking dark lateral stripe through eye to side of body; tadpole with or without dark lateral stripe but lacking light stripe above eye; **go to 2.**

2a Head lacking a dark triangular blotch or expanded figure between the eyes; body slender and with three dark dorsal stripes or series of spots; snout pointed; tadpole with dark stripe on tail musculature, no light stripe above eye, a dark chest, and no small dark spots on dorsum.

Pseudacris nigrita—**Southern Chorus Frog** . . . **page 164.**

From left to right:

Dorsal view of Southern Chorus Frog (*Pseudacris nigrita*).

Lateral view of Southern Chorus Frog (*Pseudacris nigrita*) tadpole.

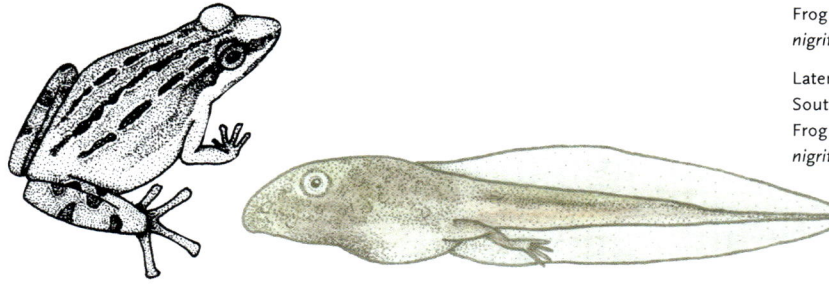

2b Head with a dark triangular blotch or expanded figure between eyes (see illustrations with couplets 3–5); body slender to stout; snout rounded (see illustrations with couplets 3–5); tadpole variously marked but never as above (see illustrations with couplets 3–5); **go to 3.**

3a Body with conspicuous dark blotches on sides and near groin; tadpole with tall caudal fin.

Pseudacris ornata—**Ornate Chorus Frog** . . . **page 167.**

From left to right:

Dorsal view of Ornate Chorus Frog (*Pseudacris ornata*).

Lateral view of Ornate Chorus Frog (*Pseudacris ornata*) tadpole.

3b Body lacking conspicuous dark blotches on sides (see illustrations with couplets 4–5); tadpole lacking noticeably tall caudal fin (see illustrations with couplets 4–5); **go to 4.**

4a Body with conspicuous dark X-shaped marking on back; toe tips with greatly expanded discs; tadpole with dark speckling on throat.

Pseudacris crucifer—**Spring Peeper** . . . **page 171**.

4b Body without dark X-shaped mark, but with dark stripes or blotches; reduced toe pads (see illustrations with couplet 5); tadpole lacking dark pigment on throat (see illustrations with couplet 5); **go to 5**.

5a Dorsum usually with two longitudinal broad dark stripes, typically in the form of reversed parentheses; tadpole with a noticeably brassy appearance and lacking spots on dorsum.

Pseudacris brachyphona—**Mountain Chorus Frog** . . . **page 175**.

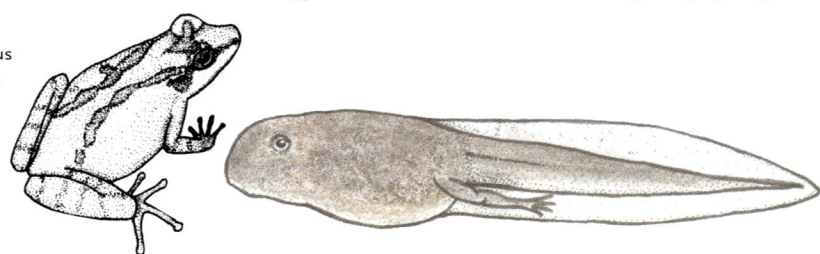

5b Dorsum usually with three longitudinal dark stripes or series of dark spots; tadpole not noticeably brassy but with small dark spots on dorsum.

Pseudacris feriarum—**Upland Chorus Frog** . . . **page 179**.

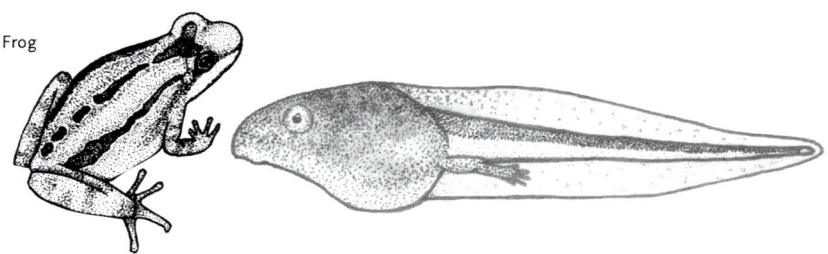

Adult Little Grass Frog (*Pseudacris ocularis*), Baker County, GA.

Little Grass Frog
Pseudacris ocularis (Bosc and Daudin, 1801)

DESCRIPTION The smallest native frog of North America, this species attains a maximum snout–vent length of about 0.7 in (18 mm). The tips of its digits are noticeably expanded, and the dorsal ground color is tan or grayish tan. A wide, dark lateral stripe is present beginning on the snout, passing through the eye, continuing in undiminished intensity to the level of the forelimbs, and gradually blending with the ground color before reaching the groin. The dorsum has a less prominent median stripe, and on most individuals two obscure dorsolateral stripes are present. The skin of the dorsum is relatively smooth. Tadpoles in this species are characterized by a distinct, wide dark stripe along the tail musculature and dark mottling on the tail fins.

Because of their small size, Little Grass Frog adults are unlikely to be confused with any other species except introduced Greenhouse Frogs (*E. planirostris*), which have digits that taper at the tip and contain no expanded toe pad. The most similar tadpole in Alabama is that of the Southern Chorus Frog (*P. nigrita*), which has a weak dark stripe along the tail musculature and clear tail fins.

ALABAMA DISTRIBUTION Little Grass Frogs are known only from the lower half of Houston County, in extreme southeastern Alabama.

HABITS This tiny frog prefers low pine flatwoods, marshes, and cypress savannas (Liner et al. 2008). During the day it frequents damp grassy

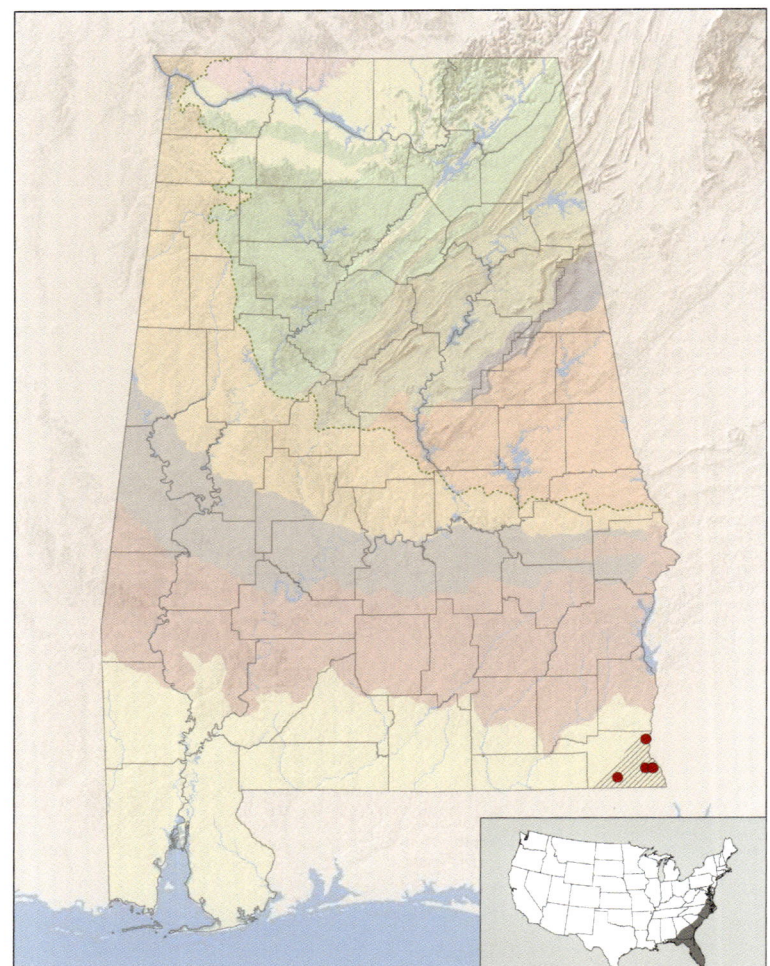

Distribution of Little Grass Frog (*Pseudacris ocularis*). The presumed range of the species in Alabama is indicated by hatching. Solid dots indicate localities of specimens or photographs examined by the authors or ADCNR/Natural Heritage Program occurrence records believed to be valid. Inset map depicts approximate range in the United States.

swales, sedges, and sphagnum along the edges of cypress ponds, where it clings to herbaceous vegetation and low shrubs. Favored breeding sites are grassy rain-filled depressions and semi-permanent ponds. Males usually perch 0.4–40 in (1–100 cm) above water on a grass stem or leaf. The call is an insect-like high-pitched tinkling or chirping, resembling the calls of some long-horned grasshoppers. The calling season is extensive, with tadpoles being present in wetlands during winter, spring, and summer months (Babbitt and Tanner 2000). Eggs, which number up to 200 in total, are laid singly or in clusters of about

25 on the bottom of shallow aquatic sites or on submerged vegetation (Wright and Wright 1949). Breeding congregations in Houston County have been recorded from January to July. Because this species breeds during the winter, it does not experience a period of inactivity during the coldest months of the year.

Diet of adults likely includes small insects. Tadpoles eat periphyton and phytoplankton that are scraped from aquatic vegetation. Tadpoles may also scavenge nutrients from dead tadpoles.

CONSERVATION AND MANAGEMENT This species is common in the Florida Panhandle and southern Georgia and continues to be detected sporadically in Houston County. The species is designated by ADCNR as a Priority 3 (species of moderate conservation concern; Shelton-Nix 2017). However, this region contains several subspecies and mitochondrial clades of amphibians and reptiles, making it of biogeographic importance. For that reason, the species is listed as critically imperiled by ANHP. Care should be made to maintain some areas in the open pine savannas that likely characterized this area in the ancestral landscape. Frequent fire will be a crucial management tool to reach this goal. The species uses temporary wetlands in agricultural pasture lands (Babbitt and Tanner 2000). Chytrid fungal disease (*Batrachochytrium dendrobatidis*) has been detected in the species (Rizkalla 2010).

TAXONOMY This species is part of the *crucifer* clade and is sister to Spring Peeper (*P. crucifer*) (Moriarty and Cannatella 2004). It has no recognized subspecific variation.

Southern Chorus Frog
Pseudacris nigrita (Le Conte, 1825)

DESCRIPTION Southern Chorus Frogs are small anurans, attaining a maximum snout–vent length of around 1.3 in (32 mm). The tips of the digits are moderately expanded in this species, and the skin is smooth to slightly pustulate. On the upper lip, a light stripe bordered above by a dark stripe is present, and these stripes pass through the eye and extend onto the sides. The dorsum is gray to brown with three dark stripes or a longitudinal series of dark spots. The median stripe terminates anteriorly between the eyes or on the snout. No dark triangular mark is present on the head. Tadpoles in this species are characterized by a faint dark stripe along the tail musculature and clear tail fins.

Adult Southern Chorus Frogs are most similar in appearance to the Upland Chorus Frog (*P. feriarum*), which possesses a triangular mark on the top of the head, is more robust, and has a more rounded snout than Southern Chorus Frogs. The most similar tadpole in Alabama is that of Little Grass Frogs (*P. ocularis*), which has a bold, wide dark stripe along the tail musculature and dark mottling on the tail fins.

ALABAMA DISTRIBUTION This species is restricted to the Lower Coastal Plain except in the area east of the Conecuh River where its range extends northward to southern Russell County.

HABITS In Alabama these frogs are the Lower Coastal Plain counterpart to the Upland Chorus Frog, except in the area east of the Conecuh

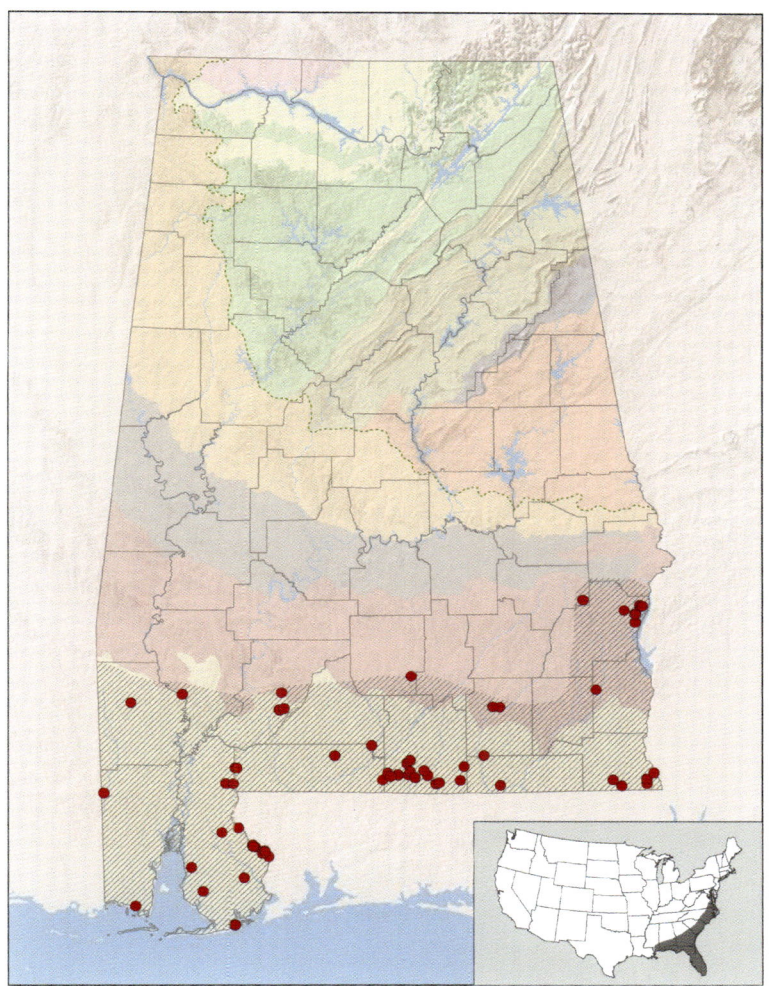

Distribution of Southern Chorus Frog (*Pseudacris nigrita*). The presumed range of the species in Alabama is indicated by hatching. Solid dots indicate localities of specimens or photographs examined by the authors or ADCNR/Natural Heritage Program occurrence records believed to be valid. Inset map depicts approximate range in the United States, with dark shading indicating the range of Southern Chorus Frog and light shading indicating the range of all other subspecies.

River where the ranges of the two species overlap. Flooded fields and roadside ditches, weedy margins of shallow flatwoods ponds, and temporary woodland pools serve as breeding sites for Southern Chorus Frogs (Liner et al. 2008), especially in areas with sandy soils. Here, males may be heard calling from late January through May. Heavy thundershower activity may initiate vigorous calling by a group of males almost anytime during summer and, on rare occasions, during fall. It is doubtful that any breeding occurs during summer and fall, however, when these frogs are thought to dig into loose soil and remain inactive. Where grassy vegetation is abundant, calling males of Southern Chorus Frogs are usually well concealed and positioned vertically, with only the head protruding above water. They are considerably more difficult to locate and are more elusive than other Chorus Frogs. The call is pulsed with a series of clicking sounds given in succession and at a decreasing rate. This creates a sound similar to rubbing a thumb slowly across a comb. Females produce up to 160 eggs, which are deposited in small clusters attached to aquatic vegetation.

Differences in habitat preference between Upland Chorus Frog and Southern Chorus Frogs are significant. In areas where the two are sympatric, Southern Chorus Frogs are likely to be found where the soil is sandy and friable, whereas Upland Chorus Frogs occur mostly in places where the soil is heavier. In addition to habitat preference, these two species avoid contact in the zone of sympatry through character displacement of calls. Within the zone of sympatry Southern Chorus Frogs retain calls that are similar to their calls outside the zone of sympatry, but males of Upland Chorus Frogs increase call pulse rate and pulse number (relative to calls of males outside of this zone), apparently to differentiate their call from those of Southern Chorus Frogs.

Diet of adults likely includes small insects. Tadpoles eat periphyton and phytoplankton that are scraped from aquatic vegetation. Tadpoles may also scavenge nutrients from dead tadpoles.

CONSERVATION AND MANAGEMENT This is a common frog along the southern tier of counties in Alabama, and it receives no regulatory protection in Alabama. Fire is the most cost-effective management tool for maintaining the habitat features selected by Southern Chorus Frogs, and this management tool, along with stand thinning, is likely to improve habitat for the species (Steen et al. 2010). These frogs are known to occupy wetlands surrounded by agricultural pastures (Babbitt and Tanner 2000).

TAXONOMY This species is part of the trilling clade and is sister to the Upland Chorus Frog (Moriarty and Cannatella 2004). Two subspecies of this frog were recognized by Mount (1975). However, Lemmon et al. (2007) found no phylogeographic structure within the species that corresponded to recognized subspecies. Based on this recent evidence, we consider this to be a single species without subspecific variation.

Adult Ornate Chorus Frog (*Pseudacris ornata*), Covington County, AL.

Ornate Chorus Frog
Pseudacris ornata (Holbrook, 1836)

DESCRIPTION Ornate Chorus Frogs are moderately small, stout anurans that attain a maximum snout–vent length of about 1.4 in (35 mm). The tips of digits are somewhat rounded but not obviously expanded. There is a dark triangular mark on the head between the eyes, and on most individuals, two irregular wide stripes or blotches on the dorsum. A dark stripe is present from the nostrils through the eye to just above the arm. These dark markings continue along the sides of the body as a sharply defined dark stripe or blotch about midway between the insertion of the arm and leg and a dark blotch or spot in the antero-dorsal region of the groin. The latter marking sometimes is connected to the dorsal stripe. The ground color is variable, ranging from green to pink or gray. The green morph results from a rare dominant allele (Blouin 1989), and the other two are created by the combined effects of genetics and environment (Travis and Trexler 1984). The region of the groin is washed with yellow or orange. The skin on the dorsum is quite smooth. Tadpoles of Ornate Chorus Frogs are notable in being about as tall as they are long, having clear tail fins, and having a purplish cast to the body.

Because of their bulky appearance, adults of Ornate Chorus Frogs are not likely to be confused with any other Chorus Frog in Alabama.

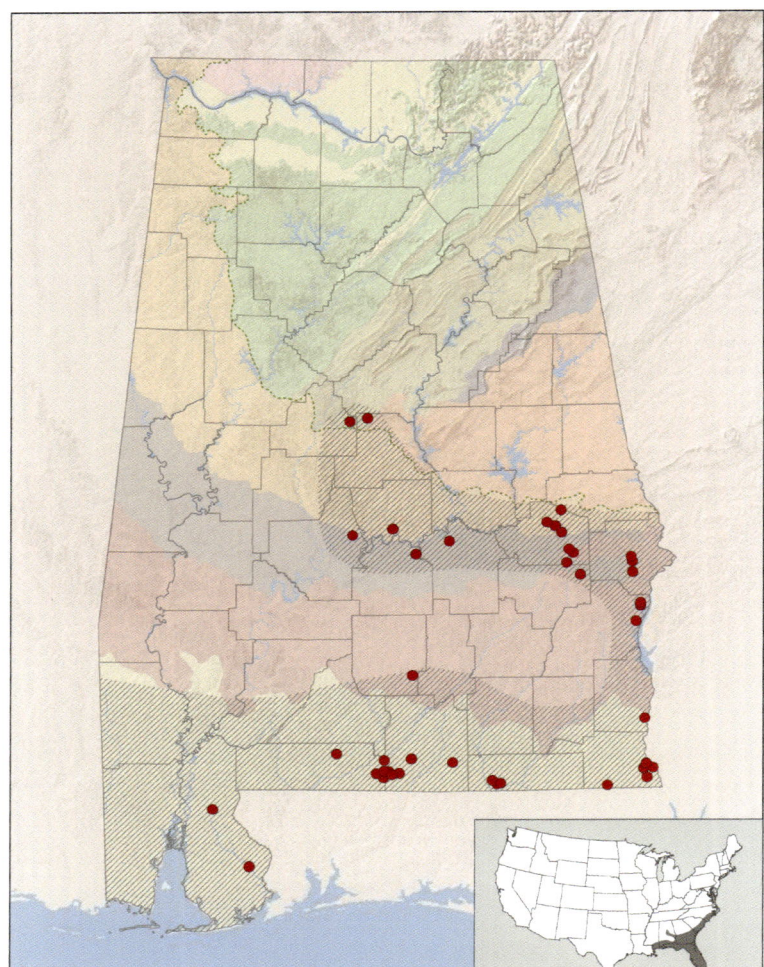

Distribution of Ornate Chorus Frog (*Pseudacris ornata*). The presumed range of the species in Alabama is indicated by hatching. Solid dots indicate localities of specimens or photographs examined by the authors or ADCNR/Natural Heritage Program occurrence records believed to be valid. Inset map depicts approximate range in the United States.

Similarly, the shape and color pattern of the tadpoles immediately distinguish them from all other *Pseudacris* tadpoles.

ALABAMA DISTRIBUTION This species is restricted to the Coastal Plain and Fall Line Hills regions, penetrating margins of the Red Hills and Black Belt, likely in areas with inclusions of sandy soils.

HABITS This frog is seldom encountered except during the breeding season, which extends from December or January, depending on rainfall, to March. However, on rare occasions, perhaps associated with drought, reproduction may take place during summer months. The earliest record of males calling in Alabama is November 15 in Escambia County. Shallow transient pools and ponds, particularly those with abundant emergent vegetation, are the chief breeding sites of the Ornate Chorus Frog. Although they often share breeding sites with other *Pseudacris*, they are less likely to breed in shallow roadside ditches and seepage areas than other Alabama Chorus Frogs. Adults migrate several hundred yards/meters to breeding sites at night during and immediately after rain showers (Todd and Winne 2006). A calling male usually sits 1–10 in (2.5–25.4 cm) above the water in clumps of grass or on floating debris. The call note, given at the rate of 65–80 per minute, is a high-pitched "peep" or "peet," somewhat like the notes of Spring Peeper (*Pseudacris crucifer*) and Oak Toads (*Anaxyrus quercicus*), but more sharply abbreviated. The note has been likened to the sound made when a steel chisel is struck with a hammer. Eggs are laid in loose clusters attached to vegetation and debris in the water and may number 25. Adult frogs are fossorial, digging refuges with the forelimbs into loose sandy soil (Brown and Means 1984). These refuges serve as the center of summer and fall activities; individuals have even been observed to bask at these burrows during periods of light snow (Ashton and Ashton 1988).

Diet of adults likely includes small insects. Tadpoles eat periphyton and phytoplankton that are scraped from aquatic vegetation. Tadpoles may also scavenge nutrients from dead tadpoles.

CONSERVATION AND MANAGEMENT Ornate Chorus Frogs are designated by ADCNR as a Priority 3 (species of moderate conservation concern; Shelton-Nix 2017) because some populations seem to have disappeared and the species is sensitive to changes in habitat quality. At the Savannah River Site, where drift fences were used to monitor anuran populations long term at Rainbow Bay, and where fire was excluded from inside the fence, allowing Sweetgum (*Liquidambar styraciflua*) to slowly dominate the overstory, a progressive loss of Ornate Chorus Frogs and increase in Marbled Salamanders (*Ambystoma opacum*) were documented (Pechmann et al. 1991). Based on this finding, fire appears to be important to the species because of its effect on retaining open habitat structure. Thus, fire is the most cost-effective management tool for maintaining habitat features appropriate for Ornate Chorus Frogs, and this tool, along with stand thinning, is likely to improve habitat for the species (Steen et al. 2010). Occupancy of this species is severely reduced in urbanized areas, making the species an indicator of healthy wetlands (Guzy et al. 2012).

TAXONOMY This species is part of the fat frog clade and is sister to *P. streckeri* + *P. illinoensis* (Moriarty and Cannatella 2004). It has no recognized subspecific variation. However, three mitochondrial lineages are known, with Alabama specimens belonging to the southern clade (Degner et al. 2010).

Adult Spring Peeper (*Pseudacris crucifer*), Marengo County, AL.

Spring Peeper
Pseudacris crucifer (Wied-Neuwied, 1838)

DESCRIPTION Spring Peepers are small frogs, attaining a maximum snout–vent length of about 1.4 in (35 mm). The tips of the digits are expanded into adhesive discs, and the dorsal skin is smooth. The dorsum has an X-shaped dark mark that is the character used to identify this species in the field. A transverse interorbital bar usually is present, but sometimes it is reduced to a medial blotch. The ground color is tan or light brown, contrasting with the dark dorsal mark. Significant geographic variation seems to be lacking among Alabama populations of this frog. Tadpoles of Spring Peeper have a more-or-less uniform color pattern to the body, tail musculature that shades gradually from dark dorsal coloration to light ventral color, and dark mottling to the tail fins.

The X-shaped mark on the dorsum of adults allows this species to be distinguished from all others. The tadpoles are most like those of Little Grass Frogs (*P. ocularis*), which also have dark mottling to the tail fins but possess a bold, wide dark stripe along the tail musculature.

ALABAMA DISTRIBUTION This species is common to abundant in every part of the state.

HABITS In the seasonal progression of anuran reproductive activity in Alabama, breeding of Spring Peepers commences a few weeks later than the other *Pseudacris*, except for Southern Chorus Frogs (*P. nigrita*), which Spring Peepers precede. The male's clear, high-pitched

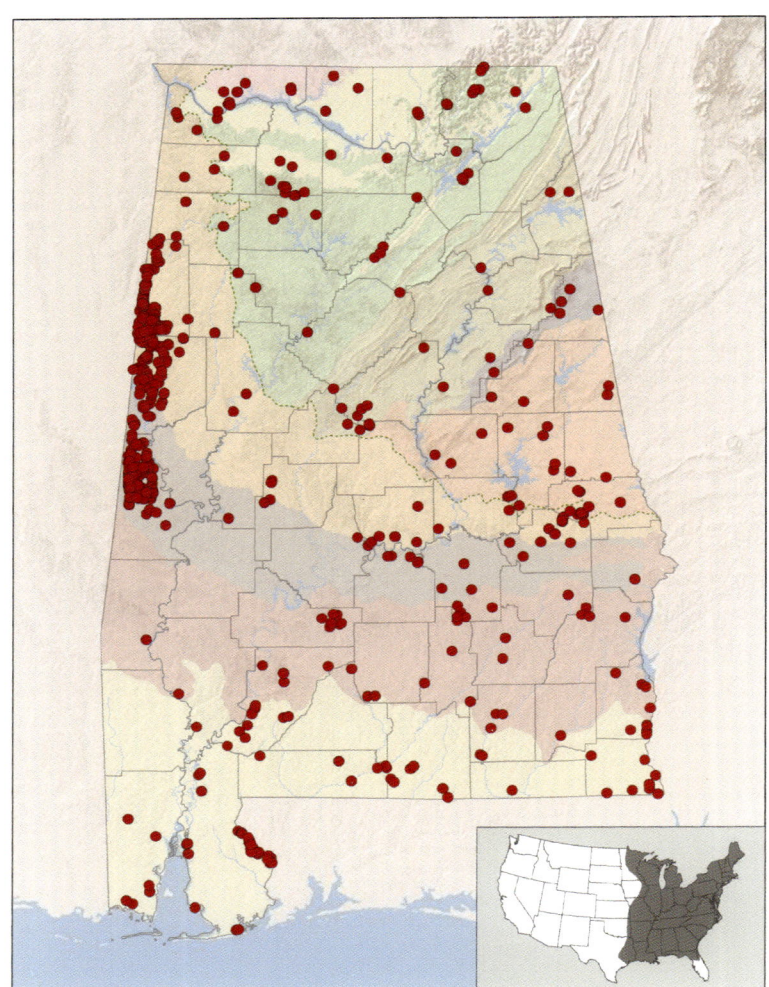

Distribution of Spring Peeper (*Pseudacris crucifer*). Solid dots indicate localities of specimens or photographs examined by the authors or ADCNR/Natural Heritage Program occurrence records believed to be valid. Inset map depicts approximate range in the United States.

whistles or peeps, rising slightly at the end and issued at a rate of about one note per second, may be heard from individual males, or small choruses of them, as early as October. Full choruses usually begin in late December in southern Alabama and in January or February in northern Alabama. Call detection increases as temperatures decrease in January and February (Steen et al. 2013).

The breeding site is most frequently a temporary or semi-permanent pond or pool, preferably one with abundant emergent vegetation. However, both open and forested wetlands can be occupied (Liner et al. 2008). Adults migrate to breeding sites at night during and immediately after rain showers (Todd and Winne 2006). Males perch from a few inches to several feet above the water on sticks, bushes, and vines, or in clumps of grass. Male choruses differ dramatically in activity from site to site, and calling may occur during both daylight and nighttime hours (Todd et al. 2003), likely being triggered by local rainfall (Kirlin et al. 2006). The female deposits several hundred eggs at a time, sticking them to vegetation under water. The tadpoles that emerge from these eggs are only marginally unpalatable to predators (Adams et al. 2011). The breeding season lasts until late April or into May. Frequently, Spring Peepers breed in company with one or more other species of *Pseudacris*. Following breeding, adults move to damp places in wooded areas where they become secretive and difficult to find. During particularly cold periods adults seek refuge in knots of tree trunks or under logs and loose bark of dead trees. Otherwise, this species alters blood chemistry to create an antifreeze that allows activity throughout the winter (Churchill and Storey 1996). During summer and fall periods, Spring Peepers occupy the same wooded wetlands used as breeding sites.

Diet of adults likely includes small insects. Tadpoles eat periphyton and phytoplankton that are scraped from aquatic vegetation. Tadpoles may also scavenge nutrients from dead tadpoles.

CONSERVATION AND MANAGEMENT Because of their wide distribution and generally large population sizes, this species receives no regulatory protection in Alabama. Spring Peepers are relatively unaffected by timber harvest (Gibbs 1998). Additionally, occupancy of this species increases in human-created wetlands that have persisted for long periods of time (Birx-Raybuck et al. 2009). Long-term calling records document that the calling season occurs progressively earlier (Walpole et al. 2012), likely associated with climate change. This species experiences mass mortality events in association with the fungal genera *Saprolegnia* and *Leptolegnia* (Ruthig 2009), and chytrid fungal pathogens have been detected in the species (Saenz et al. 2010). Ranaviruses are known in this species, and high prevalence of this pathogen has been noted in mass mortality events (Miller et al. 2011). Increased levels of environmental copper, especially in association with exposure to UV-B, reduce tadpole fitness (Baud and Beck 2005).

TAXONOMY This species is part of the *crucifer* clade and is sister to Little Grass Frogs (Moriarty and Cannatella 2004). The two subspecies of this common frog, recognized as *Hyla* by Mount (1975), do not appear as monophyletic lineages in recent phylogeographic examinations of this species. Therefore, we recognize no formal subspecies, but do follow Austin et al. (2002) in recognizing four lettered clades. Clade B is documented for Alabama by a specimen from Barbour County (Austin et al. 2002). Clade D is not documented from Alabama but is known from Murfreesboro, Tennessee, and we infer it to be present in Alabama north of the Tennessee River.

Adult Mountain Chorus
Frog (*Pseudacris brachy-
phona*), Jackson County, AL.

Mountain Chorus Frog
Pseudacris brachyphona (Cope, 1889)

DESCRIPTION Mountain Chorus Frogs are small anurans, attaining a maximum snout–vent length of around 1.4 in (35 mm). The tips of the digits are moderately expanded, and the skin is pustulate. A dark triangle is present between the eyes, the upper lip has a light stripe that extends posteriorly to the level of the tympanum, and the dorsum has a brown to grayish ground color. The dorsal ground color contrasts with a pair of outward-curving dorsolateral dark bars on the back of many individuals. However, these dark marks often are inconspicuous because they may be heavily dissected into irregular shapes or they may be faded in frogs that have lightened their color. Dorsolateral bars occasionally are absent and are replaced by small irregularly shaped dark spots or flecks. The tadpole of Mountain Chorus Frogs has tail musculature that shades gradually from the dark dorsal coloration to the light ventral color; the tail fins are clear.

Mountain Chorus Frogs are most easily confused with Upland Chorus Frogs (*P. feriarum*), a species that possesses three dark, relatively straight dorsal stripes rather than a pair of bowed stripes. The tadpole is most like that of Upland Chorus Frogs, which differs by having distinctly two-toned tail musculature (dark dorsal half and light ventral half).

Adult Mountain Chorus Frog (*Pseudacris brachyphona*), Walker County, AL.

ALABAMA DISTRIBUTION Mountain Chorus Frogs are found from within the Fall Line Hills northward to the Tennessee River Valley and the mountains north of the Tennessee River in Jackson County.

HABITS Mountain Chorus Frogs, as with other Chorus Frogs in Alabama, are seldom collected except during the breeding season, which usually begins in December or January, depending on the weather, and lasts until mid to late April. This winter activity may involve the same antifreeze blood chemistry observed in Spring Peepers (*Pseudacris crucifer*). Breeding sites are usually in hilly, wooded, or partially wooded areas and include seepages at the base of hills or mountains, flooded roadside ditches, and other shallow pools and puddles. During the height of the breeding season the mating call, a rasping "wrrink-wrrink . . . wrrink," may be heard both day and night. These frogs are shy, and they usually cease calling abruptly at the slightest disturbance. If approached, they will frequently dive to the bottom of a pool and take shelter under debris until the intruder has left. Small clumps of 10–30 eggs are deposited under water. Usually the clumps are attached to vegetation or sticks, but occasionally they are free and sink to the bottom (Brown 1956). Once breeding is complete, adults move up slope from the breeding site, occupying forested areas where this species is presumed to burrow in loose soil or leaf litter. These seasonal movements may be as much as 0.6 mi (1 km).

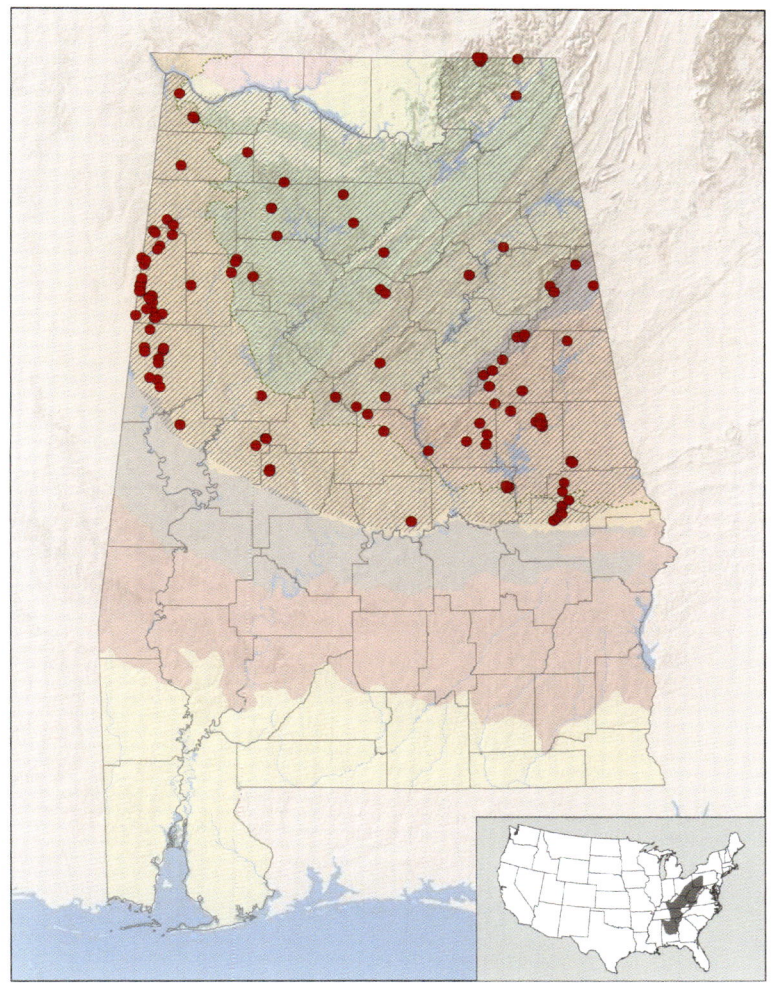

Distribution of Mountain Chorus Frog (*Pseudacris brachyphona*). The presumed range of the species in Alabama is indicated by hatching. Solid dots indicate localities of specimens or photographs examined by the authors or ADCNR/Natural Heritage Program occurrence records believed to be valid. Inset map depicts approximate range in the United States.

Diet of adults likely includes small insects. Tadpoles eat periphyton and phytoplankton that are scraped from aquatic vegetation. Tadpoles may also scavenge nutrients from dead tadpoles.

CONSERVATION AND MANAGEMENT This species is locally abundant in the northern two-thirds of the state and, because of this, receives no regulatory protection in Alabama. Mountain Chorus Frogs can be locally abundant in second-growth areas, especially where puddles of water accumulate. These frogs appear to be compatible with land-use practices associated with forestry. However, patch occupancy of this species has decreased over time in the northeast US, perhaps portending declines elsewhere (Weir et al. 2014).

TAXONOMY Lemmon et al. (2007) documented a northern and a southern clade within this species, with all Alabama specimens belonging to the southern clade. After the original manuscript for this book was accepted, Ospina et al. (2020) described each clade as an independent species, with the northern clade retaining the name Mountain Chorus Frog and the southern clade described as a new species, Collinses' Mountain Chorus Frog (*P. collinsorum*). All Alabama specimens were assigned to *P. collinsorum*, except for a single specimen of *P. brachyphona* from Jackson County. The genomic data provide strong support that the northern and southern clades represent separate lineages. However, Ospina et al. (2020) found no morphological differences between the species and sampled call data from only a single site for each species so it is unclear whether the call differences described measure differences between sites within a single species or differences between species. We retain a single species, *P. brachyphona*, for all Alabama specimens until careful characterization of call and color variables reveal two diagnosable taxa.

Adult Upland Chorus Frog (*Pseudacris feriarum*), Marengo County, AL.

Upland Chorus Frog
Pseudacris feriarum (Baird, 1854)

DESCRIPTION This small frog attains a maximum snout–vent length of about 1.4 in (35 mm). It has toe tips that are slightly expanded and skin that is smooth to weakly granular. The upper lip has a wide light stripe that passes from the tip of the snout to the level of the arm. A dark stripe also begins on the snout, passes through the eye, and continues to the side of the body. The top of the head has an expanded dark triangular area between the eyes. On the dorsum, three dark brown stripes are present, but these may be dissected into a linear series of spots. The dark stripes are offset from a light gray, tan, or brown ground color of the dorsum. The head is relatively wide, and the snout is rounded. In populations of Upland Chorus Frogs west of the Tombigbee River in Alabama, the dorsal stripes tend to be distinctly wider and are more frequently unbroken than in other populations in the state. The tadpole of Upland Chorus Frog has tail musculature that is two-toned (dark dorsal half, light ventral half) and tail fins that are clear.

This species is quite similar in appearance to Southern Chorus Frogs (*P. nigrita*), which lack a triangular mark on the top of the head, are thinner, and have a more pointed snout than Upland Chorus Frogs. The tadpole is most like that of Mountain Chorus Frogs (*P. brachyphona*), which differ by lacking the distinctly two-toned tail musculature of Upland Chorus Frogs.

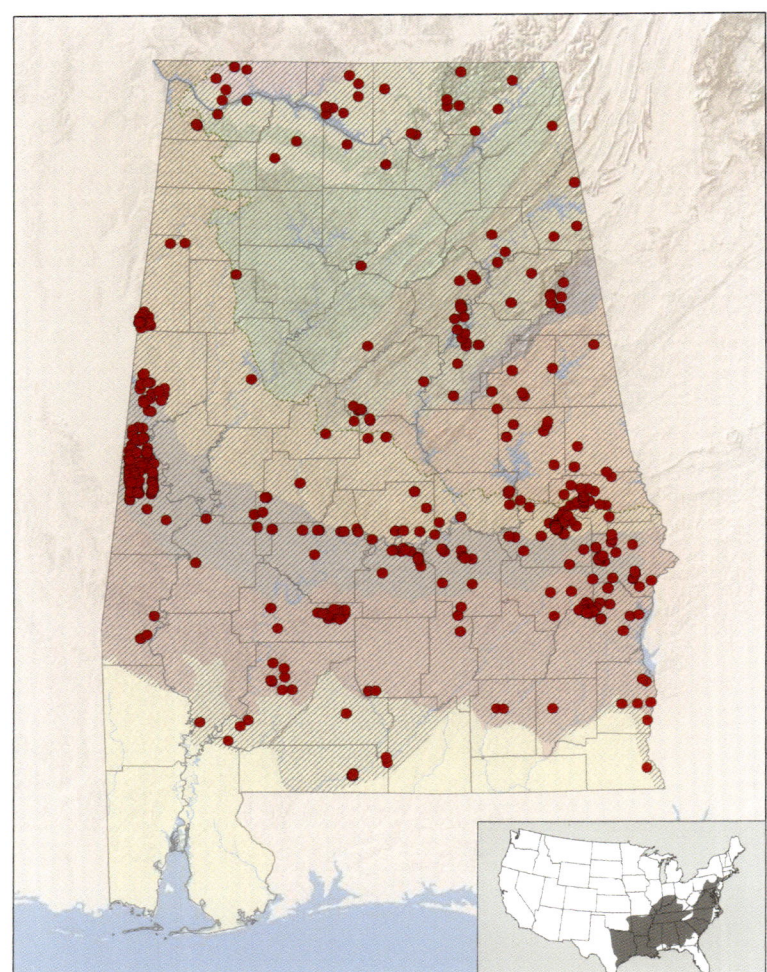

Distribution of Upland Chorus Frog (*Pseudacris feriarum*). The presumed range of the species in Alabama is indicated by hatching. Solid dots indicate localities of specimens or photographs examined by the authors or ADCNR/Natural Heritage Program occurrence records believed to be valid. Inset map depicts approximate range in the United States.

ALABAMA DISTRIBUTION The Upland Chorus Frog is common to abundant in all areas of the state southward to the southern boundary of the Red Hills region. In the Lower Coastal Plain, it is local from Clarke County eastward and apparently absent from areas west of the Tombigbee River.

HABITS The cheery sounds of this little frog, along with those of the other members of its genus, are heard in Alabama during wet weather in winter and early spring. In some years breeding may begin as early as October or November, but more often it commences around the middle of December. At this time, males begin calling from rainwater pools in ditches, fields, marshes, cypress savannas, and cypress-gum swamps (Liner et al. 2008). The call consists of a continuous series of short trills, "prreep, prreep . . . prreep," resembling the sound produced when a thumbnail is run along the teeth of a plastic comb, but somewhat more ringing. Each trill rises at the end; the pulse rate is typically faster than in the call of Southern Chorus Frogs. The call is issued as the male sits in shallow open water, on a bank, or amid clumps of grass or debris in water. These sites usually are in clay-based soils. During cloudy weather calling may continue day and night. Male choruses have distinct pulses of activity during the day but calling can be equally intense at night and differs dramatically among sites (Todd et al. 2003). These pulses of calling activity are associated with warm periods during the breeding season (Kirlin et al. 2006). Where these frogs overlap Southern Chorus Frogs in south Alabama, males increase call pulse rate and pulse number, apparently to differentiate their call from that of the more common Southern Chorus Frog; females of Upland Chorus Frogs in these areas are more attracted to males giving these modified calls than they are to calls of male Upland Chorus Frogs from the rest of their range (Lemmon 2009). During summer and fall months these frogs are found in forests, forest edges, and fields where surprisingly little is known of their activities. However, they likely bury into loose soil or leaf litter as in other members of the genus.

Eggs are laid in clusters ranging in number from about 20 to 100 each. These clusters adhere to sticks and grass. Tadpoles that emerge from these eggs are known to be palatable to fish predators (Kats et al. 1988). Breeding activity in Alabama usually ends by late April, although a cool rainy spell in summer may produce sporadic calling. During the nonbreeding season, Upland Chorus Frogs are rarely seen abroad except at night during damp weather.

Diet of adults likely includes small insects. Tadpoles eat periphyton and phytoplankton that are scraped from aquatic vegetation. Tadpoles may also scavenge nutrients from dead tadpoles.

CONSERVATION AND MANAGEMENT This species is abundant through the northern two-thirds of the state. Because of this the species receives no regulatory protection in Alabama. Upland Chorus Frogs breed in roadside ditches, and the species does well in urban settings.

Increased roads in forested areas likely increase occupancy for this species because roadside ditches provide breeding sites (Kirlin et al. 2006). This species has experienced reduced site occupancy over time in the northeastern US, perhaps portending declines elsewhere (Weir et al. 2014). Mortality events of tadpoles and juveniles, in association with high prevalence of ranavirus, have been noted for this species (Miller et al. 2011).

TAXONOMY This species is part of the trilling clade and is sister to Southern Chorus Frogs (Moriarty and Cannatella 2004). We follow Lemmon et al. (2007) in recognizing an inland and a coastal clade within this species. Only the inland clade is documented for Alabama. The type specimen for the species likely is of the coastal clade, and no obvious association of the inland clade with any previously named taxon is known (Frost et al. 2006). So, if these two clades represent separate species, the taxon in Alabama will require a new scientific name.

Holarctic Treefrogs

Genus *Dryophytes* (Laurenti, 1768)

We follow Duellman et al. (2016) in placing 20 species from North America and eastern Asia into this genus; all were previously placed in the genus *Hyla* (e.g., Faivovich et al. 2005). Seven species occur in Alabama, all of which can be distinguished from other Alabama anurans by their large toe pads and granular ventral skin. All members of the genus have males that vocalize to attract females during spring and summer months. The choruses of males typically occur in places separate from the place where the eggs are laid. In this breeding system, known as a lek, a female assesses male mates based solely upon the quality of the call given by the male. This genus is sister to *Hyla* (Duellman et al. 2016), with the ancestor of these two genera likely originating from Middle American ancestors and then crossing the Bering Land Bridge to invade Asia, Europe, and northern Africa (Frost et al. 2006).

KEY TO THE SPECIES OF *DRYOPHYTES* OF ALABAMA

1a Face with a light spot below each eye; tadpole with red saddles or red wash on tail fin and lacking an acuminate projection at the tail tip (see illustrations with couplet 2 and 4); **go to 2.**

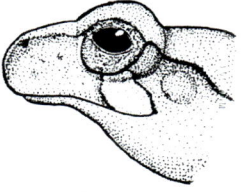

Lateral view of head of Cope's Gray Treefrog (*Dryophytes chrysoscelis*).

1b Face without a light spot below each eye (see illustrations with couplets 3–6); tadpole lacking red on tail fin or, if red, possessing a distinct acuminate projection to the tail tip or possessing a light stripe from eye to nostril (see illustrations with couplets 3–6); **go to 3.**

2a Light interspaces between dark markings on rear of thigh yellow or orange (in life); skin of dorsum warty; tadpole typically with red wash on tail fin.

Dryophytes chrysoscelis—Cope's Gray Treefrog . . . page 187.

Lateral view of Cope's Gray Treefrog (*Dryophytes chrysoscelis*) tadpole.

2b Light interspaces between dark markings on rear of thigh light green, yellowish green, or white with a turquoise cast; skin of dorsum smooth to slightly papillate or pustulate; tadpole typically with red saddles on tail.

Dryophytes avivoca—Bird-voiced Treefrog . . . page 191.

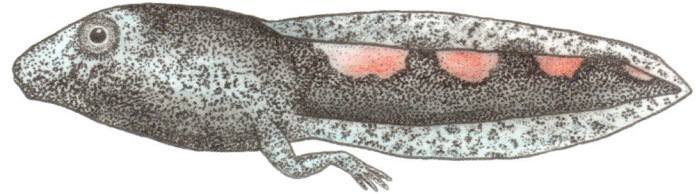

Lateral view of Bird-voiced Treefrog (*Dryophytes avivoca*) tadpole.

3a Rear of thigh with distinct light spots (yellow or orange in life; see illustrations with couplet 4); tadpole with acuminate tip or with golden-brown body (see illustrations with couplet 4); **go to 4.**

3b Rear of thigh without light spots (see illustrations with couplets 5 and 6); tadpole lacking acuminate tip or golden-brown body (see illustrations with couplets 5 and 6); **go to 5.**

4a Dorsal ground color gray, brown, or green with dark dorsal blotches; yellow spots on back of thigh; tadpole body dark brownish gray, with tall, reddish tail fin with an acuminate tip; third lower tooth row longer than width of beak.

Dryophytes femoralis—Pine Woods Treefrog . . . page 195.

Clockwise from top:

Posterior view of thigh of Pine Woods Treefrog (*Dryophytes femoralis*).

Lateral view of Pine Woods Treefrog (*Dryophytes femoralis*) tadpole.

Ventral view of Pine Woods Treefrog (*Dryophytes femoralis*) tadpole mouth parts.

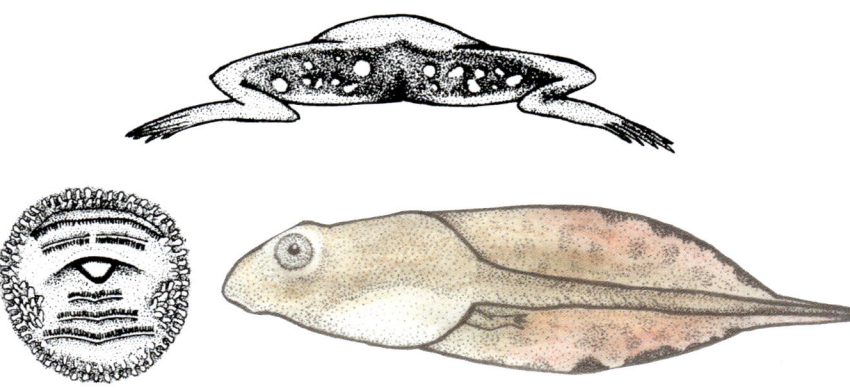

4b Dorsal ground color green with dark maroon stripe along side of body from tip of snout to midway between axilla and groin, bordered above by white or yellow; orange spots on rear of thigh; tadpole body light brown, with a tail fin possessing a light golden stripe, tail bordered above by dark stripe, and tail shape that is not noticeably tall; third lower tooth row shorter than width of beak.

Dryophytes andersonii—**Pine Barrens Treefrog . . . page 198.**

Clockwise from top:

Posterior view of thigh of Pine Barrens Treefrog (*Dryophytes andersonii*).

Lateral view of Pine Barrens Treefrog (*Dryophytes andersonii*) tadpole.

Ventral view of Pine Barrens Treefrog (*Dryophytes andersonii*) tadpole mouth parts.

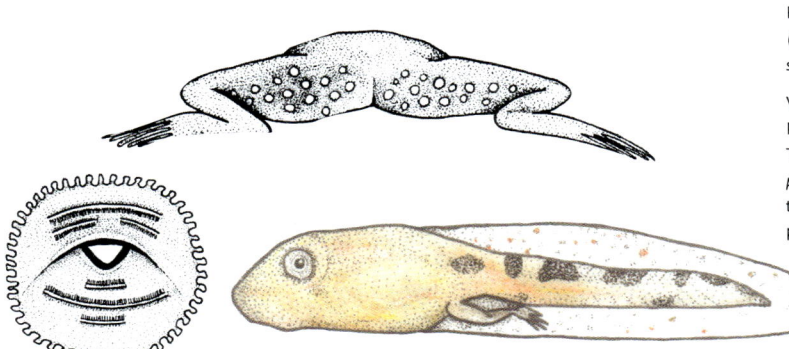

5a Adult size less than 1.6 in (40 mm) snout–vent length; pigmentation variable but never with conspicuous markings on sides; tadpole brownish gray with clear tail fin; third lower tooth row longer than width of beak.

Dryophytes squirella—**Squirrel Treefrog . . . page 202.**

Clockwise from top:

Lateral view of Squirrel Treefrog (*Dryophytes squirella*).

Lateral view of Squirrel Treefrog (*Dryophytes squirella*) tadpole.

Ventral view of Squirrel Treefrog (*Dryophytes squirella*) tadpole mouth parts.

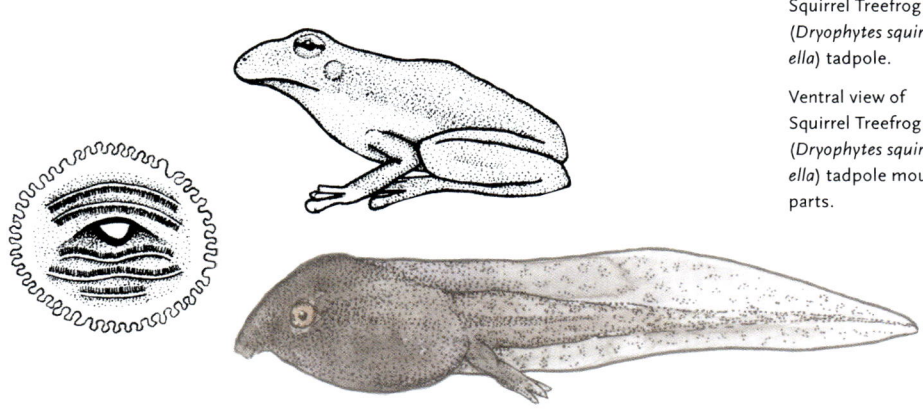

5b Adult size greater than 1.8 in (45 mm) snout–vent length; sides with conspicuous markings (see illustrations with couplet 6); tadpole variously marked but tail fin mottled or smoky gray (see illustrations with couplet 6); third lower tooth row shorter than width of beak; **go to 6.**

Ventral view of Green Treefrog (*Dryophytes cinerea*) tadpole mouth parts.

From left to right:

Lateral view of Green Treefrog (*Dryophytes cinerea*).

Lateral view of Green Treefrog (*Dryophytes cinerea*) tadpole.

6a Skin smooth; edges of light lateral stripe bold and parallel; body wall of tadpole not transparent with a light stripe from eye to nostril; tail fin mottled and not noticeably tall.

Dryophytes cinerea—**Green Treefrog . . . page 205.**

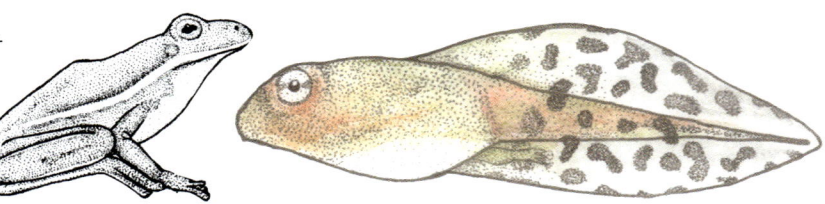

From left to right:

Lateral view of Barking Treefrog (*Dryophytes gratiosa*).

Lateral view of Barking Treefrog (*Dryophytes gratiosa*) tadpole.

6b Skin noticeably bumpy, edges of lateral light stripe scalloped or dissected; body wall of tadpole transparent with a light stripe from eye to nostril, tail fin noticeably tall, tail musculature with a smoky gray spot in center in young individuals.

Dryophytes gratiosa—**Barking Treefrog . . . page 210.**

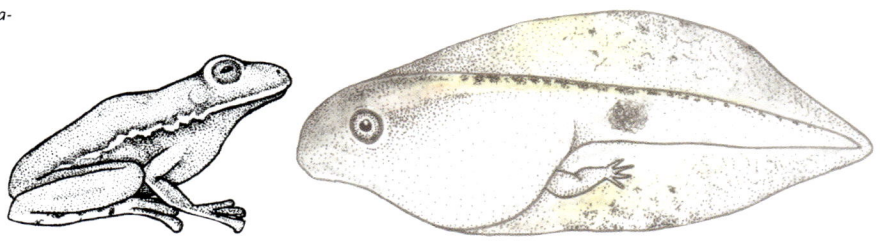

Adult Cope's Gray Treefrog (*Dryophytes chrysoscelis*), Marengo County, AL.

Cope's Gray Treefrog
Dryophytes chrysoscelis (Cope, 1880)

DESCRIPTION Cope's Gray Treefrog is a fairly large hylid that attains a maximum snout–vent length of about 2.4 in (60 mm). The tips of the digits are expanded into adhesive discs, and the dorsal skin is noticeably warty. The dorsum is light gray to dark gray or brownish (occasionally greenish) with one or two large, centrally located dark figures of varying shape. A prominent light spot usually is present below each eye. Skin of the inner surface of the thighs, and to a lesser extent along each tibia, is bright yellow or orange with dark spots and reticulations or, in some individuals, brownish with golden yellow spots. Tadpoles of Cope's Gray Treefrog have dark mottling, a red wash on the tail fin in most individuals (lacking in fishless ponds) and lack an elongate acuminate tip to the tail.

Cope's Gray Treefrog is most similar in appearance to Bird-voiced Treefrogs (*D. avivoca*) and Pine Woods Treefrog (*D. femoralis*). In Bird-voiced Treefrogs, the skin is granular, and in Pine Woods Treefrogs it is smooth (warty in Cope's Gray Treefrog). The back of the thigh is white with a turquoise cast in Bird-voiced Treefrogs and contains distinct yellow spots in Pine Woods Treefrogs (bright orange or yellow in Cope's Gray Treefrog). These three species also have tadpoles that are similar in appearance, but Bird-voiced Treefrogs differ in having bold red saddles across the tail musculature, and Pine Woods Treefrogs differ in having an acuminate tip to the tail and lacking dark mottling to the tail fin.

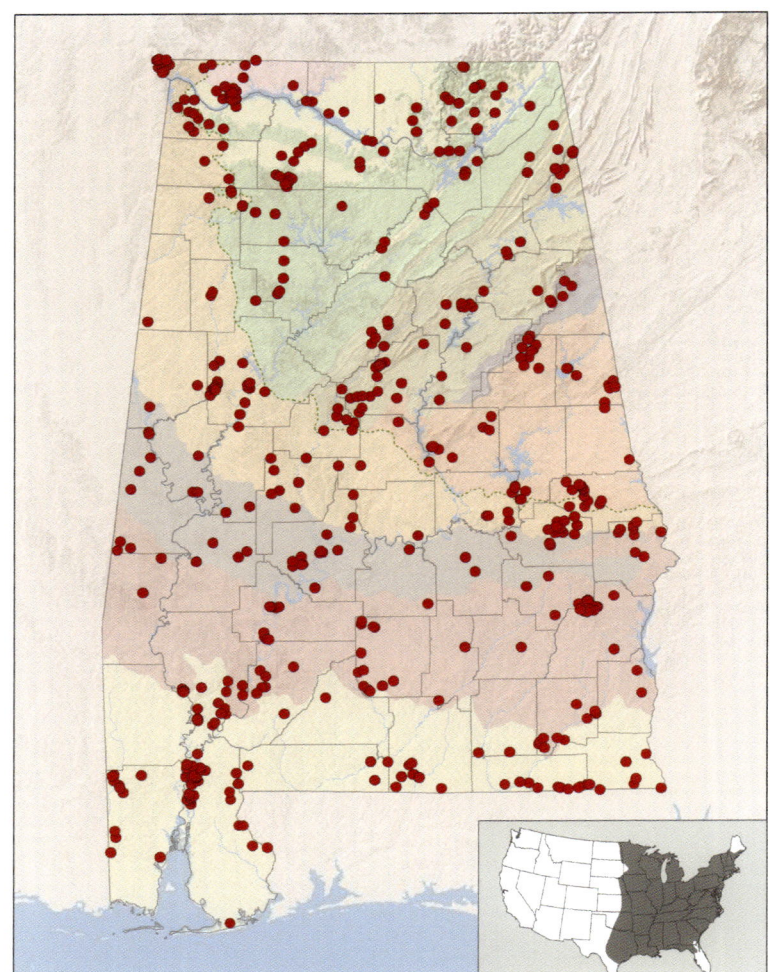

Distribution of Cope's Gray Treefrog (*Dryophytes chrysoscelis*). Solid dots indicate localities of specimens or photographs examined by the authors or ADCNR/ Natural Heritage Program occurrence records believed to be valid. Inset map depicts approximate range in the United States.

ALABAMA DISTRIBUTION Cope's Gray Treefrogs are found throughout the state.

HABITS Cope's Gray Treefrogs are common but are seldom encountered in abundance except during breeding. The breeding period extends from late March to August, with the greatest activity occurring during warm, rainy periods (Steen et al. 2013). The mating call is a short, loud trill, repeated at regular intervals. Males also announce their presence by sporadic calling from perches in trees away from breeding sites and outside the breeding season. Cope's Gray Treefrogs use a variety of aquatic sites for breeding, but streams and large bodies of water are usually shunned. The ideal breeding site appears to be a temporary or semi-permanent pool or pond, such as a flooded borrow pit, but marshes, cypress savannas, and cypress-gum swamps frequently are used (Liner et al. 2008). Males call sporadically throughout the breeding season, and calling activity is most intense from dusk to midnight (Bridges and Dorcas 2000). Males call from the ground and attract females who select males with the longest call durations, fastest pulse numbers, and fastest call rates (Gerhardt et al. 1996; Welch et al. 1998). Once a male amplexes a female, she selects a nest site in fish-free wetlands (Binckley and Resetarits 2003) where several packets, each of about 40 eggs, are laid as a surface film that may be attached to vegetation; these eggs are fertilized externally by the male. Tadpoles of this species are known to be palatable to fish predators (Kats et al. 1988), a feature that appears to drive adult females to select fish-free nest sites. After breeding, adults move to trees surrounding breeding sites where they may give short calls associated with approaching rainstorms. They may also occupy hollowed snags or rotten logs. During daylight hours they remain inactive but may emerge at night to feed. During winter adults burrow into loose soil or leaf litter at the base of trees or occupy the center of rotting logs.

Adults eat arboreal insects, and tadpoles consume algae and scrape aquatic vegetation. During dry periods Cope's Gray Treefrogs are likely to be found well away from water and may take shelter in knotholes in trees and in other such protected arboreal microhabitats. At night they may leave the trees and move to the ground to feed. During winter these frogs survive freezing temperatures by increasing glucose in the blood, creating an antifreeze (Costanzo et al. 1992).

Tadpoles in habitats populated with dragonfly predators have taller tail fins, and the tail is reddish yellow with black mottling, features that allow the tadpoles to swim faster in order to avoid the predator and direct attacks by the dragonfly to the tail (McCollum and Leimberger 1997). Such predators are so common in Alabama wetlands that nearly all tadpoles have these features.

CONSERVATION AND MANAGEMENT Cope's Gray Treefrogs are abundant and have apparently viable populations in areas such as city and county parks, where habitat has been modified heavily by human activities. For this reason, these frogs receive no regulatory protection in

Alabama. Fish must not be introduced to wetlands where maintenance of Cope's Gray Treefrog is a high priority, because these predators will consume all frog eggs and tadpoles. No other specialized management appears to be required for these frogs. Calling activities are sensitive to road density, suggesting that the species is less common in urban areas (Cosentino et al. 2014). However, the species is known to be an early colonist of human-created wetlands (Birx-Raybuck et al. 2009). Mortality of tadpoles and juveniles in association with high prevalence of ranavirus is known for the species (Miller et al. 2011). Chytrid fungal disease (*Batrachochytrium dendrobatidis*) occurs in the species and causes reduced mass of juveniles at metamorphosis and increased time to metamorphosis. Increased levels of copper also cause increased time to metamorphosis, and this time is further lengthened in tadpoles with both increased copper and infection with chytrid fungus (*Batrachochytrium dendrobatidis*; Parris and Baud 2004).

TAXONOMY No subspecies are recognized within this species. Based on the nuclear and mitochondrial genomes, Holloway et al. (2006) show a single lineage for this species. The Gray Treefrog (*D. versicolor*), a sister species long thought to be present in the state (Mount 1975), is distributed well north of Alabama (Holloway et al. 2006). These two species are part of a lineage that includes other bark-colored treefrogs (Pine Woods Treefrogs and Bird-voiced Treefrogs) plus Pine Barrens Treefrogs (*D. andersonii*) (Faivovich et al. 2005).

Adult Bird-voiced Treefrog (*Dryophytes avivoca*), Telfair County, GA.

Bird-voiced Treefrog
Dryophytes avivoca (Viosca, 1928)

DESCRIPTION This is a medium-sized treefrog, attaining a maximum snout–vent length of about 2 in (50 mm). The tips of the digits all have adhesive discs, and the dorsal skin is granular. The dorsum of adults is either grayish brown or light green, usually with an irregularly shaped large dark blotch. Young frogs are light green, lacking the blotch. A prominent light spot is present under each eye. The inner surfaces of the thighs, and to a lesser extent the tibiae, are light green, yellowish green, or white with a turquoise cast; in rare individuals the thighs will be off-white with dark spots and reticulations. The tadpoles are a stunning mottled bluish black with bold reddish saddles across the tail musculature in most individuals (some individuals lacking red).

At first glance, this species is difficult to distinguish from Cope's Gray Treefrog (*D. chrysoscelis*). However, Cope's Gray Treefrog has a warty dorsal skin (granular in Bird-voiced Treefrogs) and has bright orange on the posterior surface of the thigh (light green, yellowish green, or white with a turquoise cast in Bird-voiced Treefrogs). The tadpoles of Cope's Gray Treefrogs have reddish mottling on the tail fin but lack red saddles across the tail musculature.

ALABAMA DISTRIBUTION Bird-voiced Treefrogs are found in every regions of the Coastal Plain and occur in isolated populations above the Fall Line in the Ridge and Valley and Appalachian Plateaus regions.

Adult Bird-voiced Treefrog (*Dryophytes avivoca*), Liberty County, FL.

HABITS Bird-voiced Treefrogs are seldom encountered except during the breeding season, when males produce their distinctive calls and announce their presence. The call is a series of clear, whistling notes, given at a rate of 2–5 per second, and sounds like a person whistling to a dog. Oak Toads and Spring Peepers (*Pseudacris crucifer*) also have similar whistling notes, but theirs are not so rapidly enunciated. Calling detection is increased during spring evenings that are relatively cool (Steen et al. 2013). Neighboring males alter their call cadence to create duets in which each male interdigitates its call with its neighbor. Females prefer individual males that produce long calls and those that duet (Martínez-Rivera and Gerhardt 2008). Breeding sites include semi-permanent and permanent pools in wooded and partially wooded situations. Floodplain pools are especially favored, particularly those with an abundance of shrubs and other low, woody vegetation growing in them. Males call from perches on a limb or vine 1.5–8 ft (0.5–2.5 m) above water. Females approach and select males on these calling sites, and this is where amplexus takes place. Females then choose a nest site in water, where up to 800 eggs are deposited in packets submerged under water (Hellman 1953). Breeding takes place from April through July and is greatly enhanced by heavy rains. During the rest of the summer and fall months adults can be found in shrubs and trees surrounding the breeding site. During winter months adults remain under logs or in tree crevices at the breeding site.

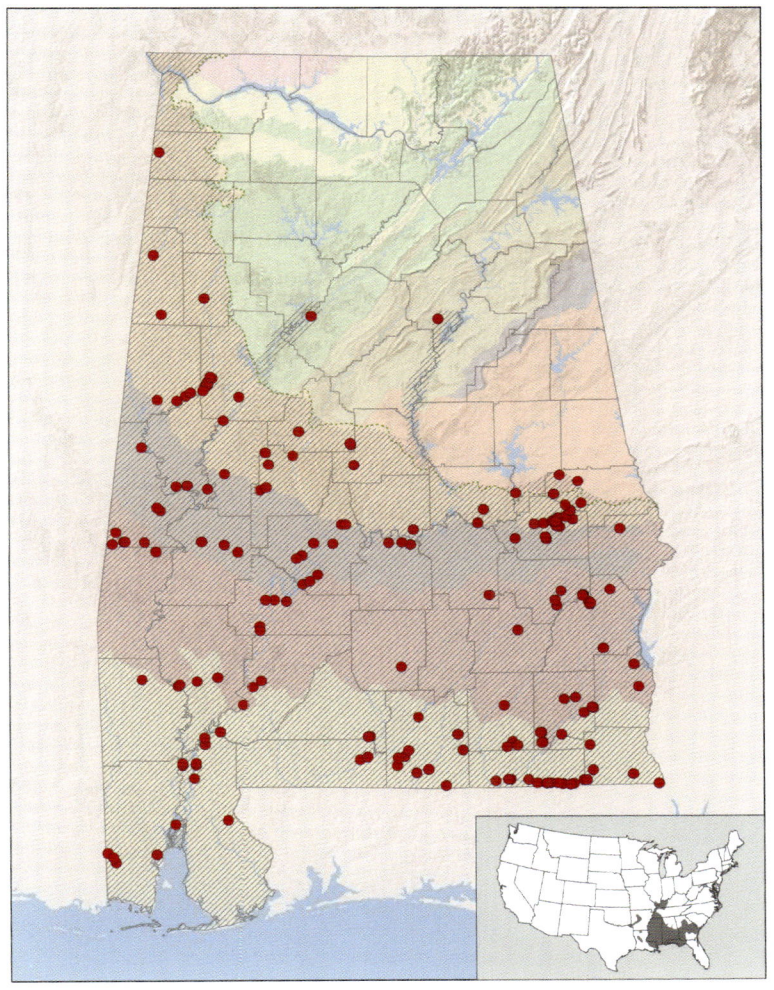

Distribution of Bird-voiced Treefrog (*Dryophytes avivoca*). The presumed range of the species in Alabama is indicated by hatching. Solid dots indicate localities of specimens or photographs examined by the authors or ADCNR/Natural Heritage Program occurrence records believed to be valid. Inset map depicts approximate range in the United States.

Diet of adults likely includes insects and other arthropods. Tadpoles eat periphyton and phytoplankton that are scraped from aquatic vegetation. Tadpoles may also scavenge nutrients from dead tadpoles.

CONSERVATION AND MANAGEMENT This species is abundant and has apparently viable populations in areas such as city and county parks, where habitat has been modified heavily by human activities. For this reason, Bird-voiced Treefrogs receive no regulatory protection in Alabama. Management efforts that maintain shrubby margins to forested wetlands should benefit these frogs.

TAXONOMY We accept Mount's (1975) argument that this is a single species with no subspecific variation. Bird-voiced Treefrogs are sister to Cope's Gray Treefrog (Holloway et al. 2006) and share many features with that species, especially a color pattern designed to mimic tree bark. Variation in the mitochondrial and nuclear genomes does not indicate consistent subspecific variation within the species (Holloway et al. 2006).

Adult Pinewoods Treefrog
(*Dryophytes femoralis*),
Baldwin County, AL.

Pine Woods Treefrog
Dryophytes femoralis (Bosc, 1800)

DESCRIPTION This species is a small treefrog, attaining a maximum snout–vent length of about 1.6 in (40 mm). The tips of the digits are expanded into adhesive discs, and the skin is smooth or slightly granular. The dorsum is gray or brown with one or more irregularly shaped dark blotches or, rarely, unicolorous. The backs of the thighs have rounded orange or yellow spots. No light spots are present below the eyes. The tadpole has a reddish wash to the tail musculature and fins, lacks dark blotching on the tail fin, and has a long, acuminate tail tip.

This species is most similar in appearance to Bird-voiced Treefrog (*D. avivoca*) and Cope's Gray Treefrog (*D. chrysoscelis*). However, these two species have a distinct, squarish white mark below the eye that is not present in Pine Woods Treefrogs. Tadpoles are most similar in appearance to those of Green Treefrogs (*D. cinerea*), but that species has large black tail spotting and lacks an acuminate tip.

ALABAMA DISTRIBUTION Pine Woods Treefrogs are common to abundant in the Lower Coastal Plain, becoming spotty in their distribution in the other Coastal Plain provinces. The species is found as far west as Marengo County in the Red Hills, Dallas County in the Black Belt, and Tuscaloosa County in the Fall Line Hills. The species penetrates the Ridge and Valley as far north as Shelby County.

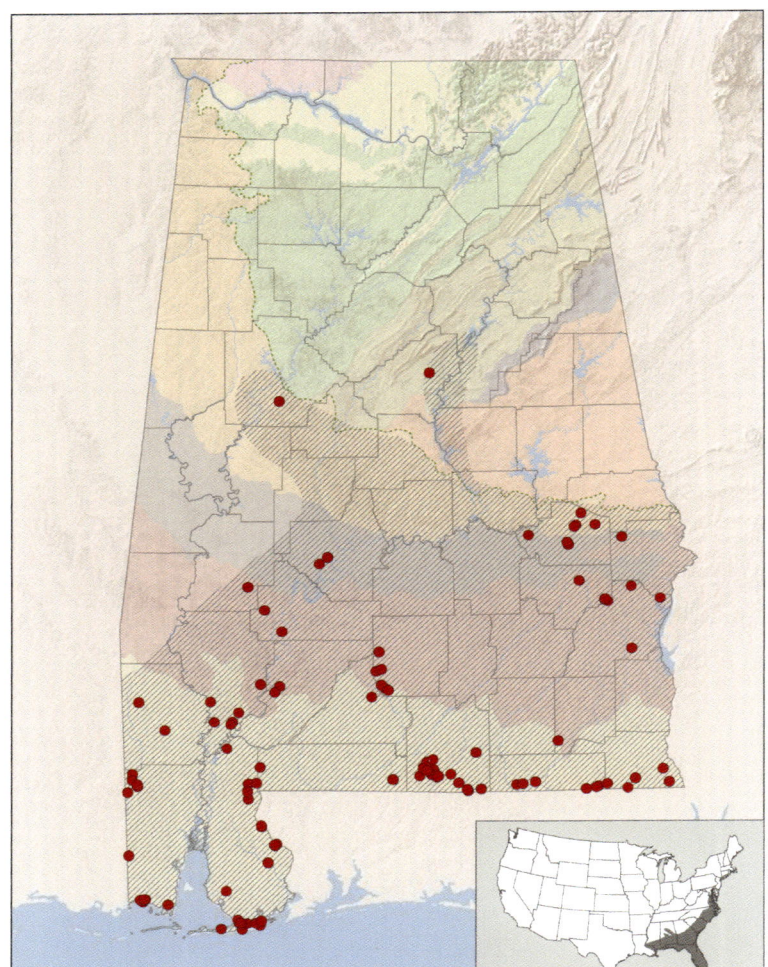

Distribution of Pine Woods Treefrog (*Dryophytes femoralis*). The presumed range of the species in Alabama is indicated by hatching. Solid dots indicate localities of specimens or photographs examined by the authors or ADCNR/Natural Heritage Program occurrence records believed to be valid. Inset map depicts approximate range in the United States.

HABITS This treefrog reportedly spends most of its time during the warm months high up in pine trees adjacent to breeding sites. Occasionally it is collected from under the bark of rotting pine stumps and logs and in the foliage of shrubs and bushes. In winter months and during hot, dry spells in summer, aggregations of Pine Woods Treefrogs are sometimes found in damp places under logs and in rotting stumps.

The breeding season in Alabama usually begins in April and extends into August. Rainy weather conjoined with warm temperatures elicits the breeding response. The usual sites occupied by this species are flooded roadside ditches, marshes, cypress savannas, cypress-gum swamps, and ponds (Liner et al. 2008). Typically, males call from perches in a tuft of grass or on a stick or limb 0.5–6 ft (0.1–2 m) above water. The call is a prolonged, monotonous "tek-a-tek-a-tek-a . . . tek," with the notes given in rapid succession. Calling peaks on warm nights during the reproductive season (Steen et al. 2013). Frequently the Pine Woods Treefrog shares a breeding site with Squirrel Treefrog (*Dryophytes squirella*), Oak Toads (*Anaxyrus quercicus*), and Eastern Narrow-mouthed Toads (*Gastrophryne carolinensis*).

The eggs are bicolored, brown and yellowish, and are laid in groups of about 100. They may be attached to vegetation or roots just beneath the surface or may form a surface film (Wright and Wright 1949). Females avoid depositing eggs in wetlands occupied by predatory fishes (Rieger et al. 2004). The presence of Pine Woods Treefrogs is an indicator of healthy ephemeral wetlands (Guzy et al. 2012). These wetlands typically are surrounded by mature pine forest habitat (Delis et al. 1996). Tadpoles emerging from fertilized eggs develop smaller bodies with shorter and wider tails and have reddish coloration when occurring in wetlands occupied by dragonfly larvae, an important predator of tadpoles (LaFiandra and Babbitt 2004).

Diet of adults likely includes insects and other arthropods. Tadpoles eat periphyton and phytoplankton that are scraped from aquatic vegetation. Tadpoles may also scavenge nutrients from dead tadpoles.

Pine Woods Treefrogs are known to hybridize with Pine Barrens Treefrogs (*D. andersonii*), producing infertile offspring (Anderson and Moler 1986).

CONSERVATION AND MANAGEMENT Pine Woods Treefrogs are abundant in their native habitat but tend to be absent from urbanized areas (Guzy et al. 2012). Agricultural ponds may be occupied, but only if they lack predatory fishes and are near pine forest remnants (Babbitt et al. 2009). The species receives no regulatory protection in Alabama. Habitat modification that thins dense pine stands and encourages growth of a grass-dominated understory along with shrubs adapted to aquatic margins should enhance populations of these frogs.

TAXONOMY No subspecies are recognized for Pine Woods Treefrogs. This species has an unresolved relationship with the *versicolor* and *eximia* groups of Faivovich et al. (2005). However, given the tree-bark color pattern of adults and the reddish coloration of the tadpole, we consider this species to be part of the *versicolor* group.

Adult Pine Barrens Treefrog (*Dryophytes andersonii*), Santa Rosa County, FL.

Pine Barrens Treefrog
Dryophytes andersonii (Baird, 1854)

DESCRIPTION Pine Barrens Treefrogs are midsized hylids, attaining a maximum snout–vent length of about 2.6 in (65 mm). The tips of the digits have enlarged adhesive discs. The dorsum of adults is green. A prominent maroon eye mask starts from the external naris and extends on each side through the eye to the middle of the side of the body. The posterior surface of the thigh has bold orange or yellow spots. The venter is uniform white. Tadpoles of Pine Barrens Treefrogs are golden brown in color with a series of bold dark bars along the tail musculature.

Pine Barrens Treefrogs are difficult to confuse with any other frog but are most similar to Green Treefrogs (*Dryophytes cinerea*). However, Green Treefrogs possess a bold white stripe along the side of the body that immediately distinguishes this species from Pine Barrens Treefrogs. The tadpoles of Pine Barrens Treefrogs are most like those of Oak Toads (*Anaxyrus quercicus*), which are found in temporary pools of water not favored by Pine Barrens Treefrogs.

ALABAMA DISTRIBUTION This species was first recorded in Alabama in 1979 by Paul Moler (1981). It is now known from Covington, Escambia, and Geneva Counties of extreme southern Alabama.

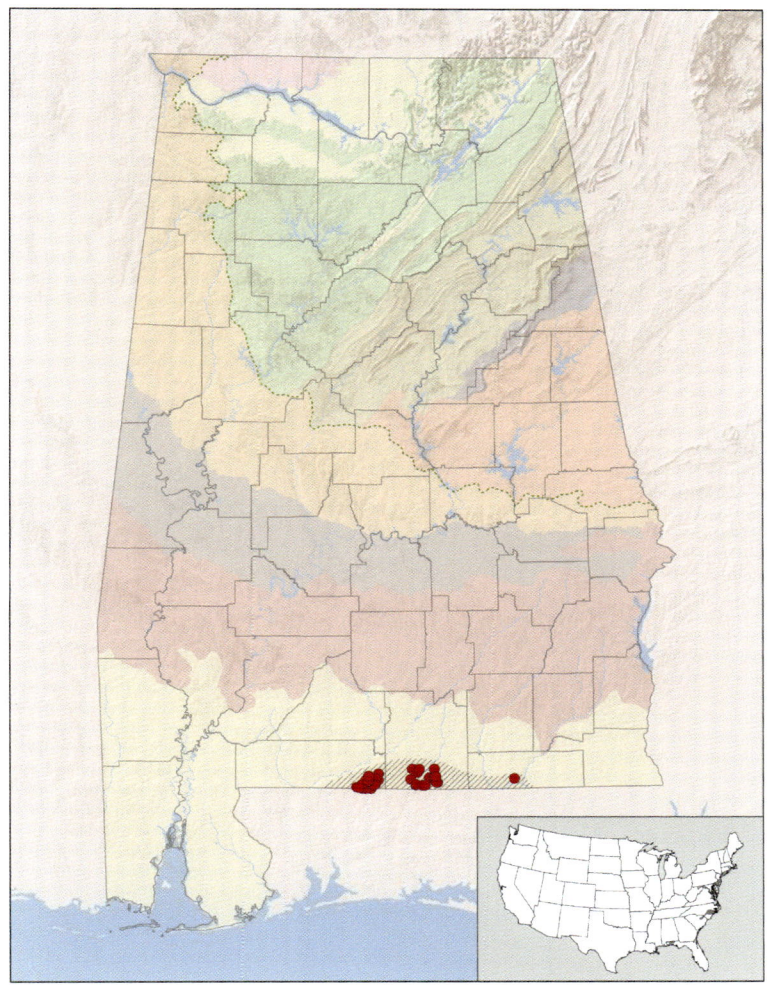

Distribution of Pine Barrens Treefrog (*Dryophytes andersonii*). The presumed range of the species in Alabama is indicated by hatching. Solid dots indicate localities of specimens or photographs examined by the authors or ADCNR/Natural Heritage Program occurrence records believed to be valid. Inset map depicts approximate range in the United States.

HABITS Males of Pine Barrens Treefrogs typically call from evergreens, such as Sweetbay Magnolia (*Magnolia virginiana*) and titi, in the shrubby, wetter portions of White-topped Pitcher Plant (*Saracenia leucophylla*) bogs. Choruses begin in early March and extend through early September. The call is like that of the Green Treefrog but is higher in pitch and delivered at a faster pace. Groups of calling males are relatively small in number, and chorus sites can be widely distributed. Calling commences at dusk and peaks approximately one hour later, diminishing rapidly after peaking. Moonlight has a negative influence on calling activity, and rainfall has a positive influence on this variable (Godwin 1985). A female selects a mate from among calling males and allows him to amplex her and then deposits up to 1,000 eggs in the acidic waters of pitcher plant seepages. During fall months occupied habitat is assumed to be the same as that during the calling season. Overwinter sites are unknown for this species except for a single observation of an individual under bark of a dead tree.

Calling choruses are exceptionally sporadic in their calling patterns, so surveys of calling sites to monitor population persistence are difficult (Means and Moler 1979). Pine Barrens Treefrogs are known to hybridize with Green Treefrogs (*D. cinerea*) and Pine Woods Treefrogs (*D. femoralis*), producing infertile offspring (Anderson and Moler 1986). Female Green Treefrogs select calls of conspecific males if both conspecific and heterospecific calls are present, regardless of the relative intensity of the two calls. However, females of Pine Barrens Treefrogs are attracted to calls of male Green Treefrogs when this heterospecific call is louder than that of a conspecific male (Gerhardt 1974).

Diet of adults likely includes insects and other arthropods. Tadpoles eat periphyton and phytoplankton that are scraped from aquatic vegetation. Tadpoles may also scavenge nutrients from dead tadpoles.

CONSERVATION AND MANAGEMENT Pine Barrens Treefrogs are found in only a few localities in Alabama, all with small populations. For this reason, the species is recognized by ADCNR as a Priority 1 (species of highest conservation concern) and is protected by Alabama's Nongame Regulation (Shelton-Nix 2017). The species is also distributed patchily across the Florida Panhandle, and so Alabama populations are an important component of the disjunct Gulf Coast segment of the species' overall geographic distribution. For this reason, it is illegal to collect this species within Alabama without a special permit. The majority of known occupied sites are located within the Conecuh National Forest, where 25 bogs have been documented to possess calling males. Unfortunately, only five of these had calling males in the most recent survey of these sites (Guyer et al. 2007). Although this may be the result of the severe drought that occurred during the survey, habitat alteration likely also affected these populations. Pitcher plant bogs require frequent low-intensity fires to allow them to maintain the open aspect and lush vegetation of these wetlands. Pine Barrens Treefrogs likely increase in abundance in fire-maintained bogs and calling activity likely is more predictable at such open sites. Additionally, bogs that are managed with frequent fire lack the thick shrub layer that allows Green Treefrogs to invade these sites and create problems of hybridization with Pine Barrens Treefrogs.

TAXONOMY We follow Warwick et al. (2015) in considering the three disjunction regions occupied by these frogs to represent a single species. Despite a general appearance that would suggest a taxonomic affinity with Alabama's other green hylids, this species shares a phylogenetic history with Alabama's blotched (bark-colored) treefrogs (Cope's Gray Treefrogs [*D. chrysoscelis*], Bird-voiced Treefrogs [*D. avivoca*], and Pine Woods Treefrogs; Faivovich et al. 2005; Wiens et al. 2006; Holloway et al. 2006). This lineage includes species from western North America and East Asia. The bulk of the range of the Pine Barrens Treefrog occurs along the Atlantic Coastal Plain of the eastern United States, but isolated populations are found along the Gulf Coastal Plain. Until its rather late discovery in the Florida Panhandle (Christman 1970), the Pine Barrens Treefrog was known only from New Jersey and the Carolinas. This is considered to be a species without subspecific variation.

Squirrel Treefrog
Dryophytes squirella (Bosc, 1800)

DESCRIPTION Squirrel Treefrogs are small, attaining a maximum snout–vent length of around 1.8 in (45 mm). Like all members of its genus, this species has expanded adhesive discs at the tips of the digits. Color patterns are extremely variable, making identification of this species difficult. The dorsal ground color may range from green to brown but will lack any sharply defined markings. However, the dorsum may be covered with inconspicuous dark flecks or small spots. A light lateral stripe often is present, but the borders of this stripe will be diffuse, never being as distinct as the bold lateral stripes in Green Treefrogs (*D. cinerea*). The rear of the thighs usually lacks any marks or contrasting colors. Most tadpoles are dark brown in ground color, but some may be black or dark green. All tadpoles have a yellow belly with a dark center and tail fins that lack markings, features that distinguish these larvae from tadpoles of all other members of *Dryophytes* in Alabama.

ALABAMA DISTRIBUTION This species is found across the Lower Coastal Plain, Red Hills, and Black Belt. It occupies the Fall Line Hills as far west as Tuscaloosa County and extends into the Ridge and Valley region as far northeastward as St. Clair and Talladega Counties.

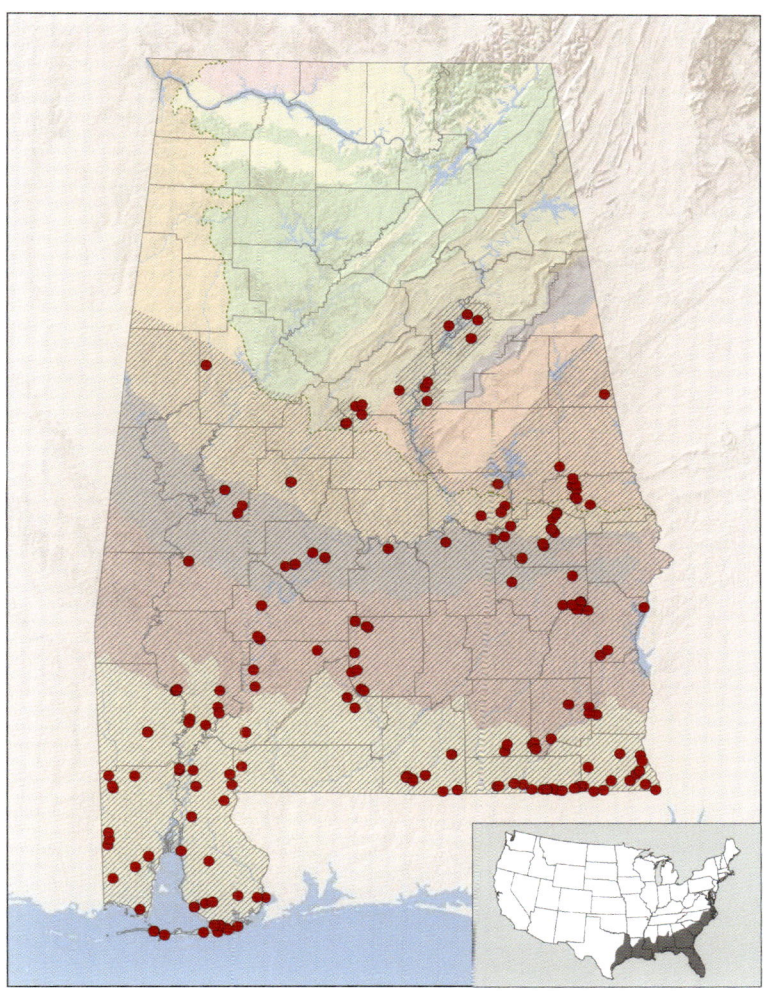

Distribution of Squirrel Treefrog (*Dryophytes squirella*). The presumed range of the species in Alabama is indicated by hatching. Solid dots indicate localities of specimens or photographs examined by the authors or ADCNR/Natural Heritage Program occurrence records believed to be valid. Inset map depicts approximate range in the United States.

HABITS In some of the low country of southern Alabama, Squirrel Treefrogs are exceedingly abundant. On warm nights following heavy rains this species is often seen in large numbers hopping on roadways and around dwellings, service stations, restaurants, and other well-lit places. During the day Squirrel Treefrogs often herald the approach of wet weather by calling intermittently from a perch in a tree or shrub. The colloquial name "rain frog," frequently applied to this species, seems appropriate. The species does well in urban settings and frequently roosts communally under facing boards of houses, emerging at night to feed at lights. Similarly, this species overwinters communally in such sites.

In Alabama this species usually breeds from about mid-April to mid-October, usually following rains. Breeding sites for Squirrel Treefrog include flooded roadside ditches, flatwoods ponds, marshes, cypress savannas, cypress-gum swamps, and small semi-permanent stock-watering ponds (Liner et al. 2008). Males usually call from a perch, 1–2 ft (0.3–0.7 m) above water, or from a bank near water's edge. The call note is a harsh, somewhat nasal rasp or "quack," repeated at a rate of about 100 times per minute. In large choruses, these calls will create a continuous rasping din. Females select males with faster call rates and are attracted to large males (Traylor et al. 2007). Up to 1,000 eggs may be produced by a female and these are deposited singly or in small groups that sink to the bottom or are attached to vegetation (Wright and Wright 1949). Females lay fewer eggs in wetlands with sunfish than wetlands lacking these predators (Binckley and Resetarits 2002).

Diet of adults likely includes insects and other arthropods. Tadpoles eat periphyton and phytoplankton that are scraped from aquatic vegetation. Tadpoles may also scavenge nutrients from dead tadpoles.

CONSERVATION AND MANAGEMENT Squirrel Treefrogs are abundant and have apparently viable populations in areas such as cities, county parks, and agricultural wetlands where habitat has been modified heavily by human activities (Babbitt and Tanner 2000, Guzy et al. 2012). In fact, occupancy of this species increases in association with agricultural lands surrounding breeding wetlands (Alix et al. 2014a), and oak forest and urbanization appear to assist gene flow among populations of Squirrel Treefrogs (Hether and Hoffman 2012). For this reason, this species receives no regulatory protection in Alabama, and no special management activities appear required to maintain these frogs in the state.

TAXONOMY This species is part of the *cinerea* group, being basal to *D. cinerea* + *D. gratiosa* (Faivovich et al. 2005). No subspecies are recognized within this species.

Adult Green Treefrog (*Dryophytes cinerea*), Limestone County, AL.

Green Treefrog
Dryophytes cinerea (Schneider, 1799)

DESCRIPTION Green Treefrogs are rather large but slender hylids that attain a maximum snout–vent length of about 2.6 in (65 mm). The tips of the digits are expanded into adhesive discs, and the dorsum is exceptionally smooth. The dorsum is light to dark green, usually with a few randomly dispersed gold flecks or small yellow spots. A lateral light stripe is present along the side of the body from the upper lip, passing below the tympanum and extending to near the groin in most individuals. In unusual cases the lateral stripe may extend only to immediately posterior to the axilla. The width of this lateral stripe differs among individuals, but the dorsal and ventral borders of this stripe are parallel, continuous, and boldly differentiated from the green dorsal color. A longitudinal light stripe also is present along the lower leg and foot. Tadpoles of Green Treefrogs have a light line from the eye to the nostril, typically have a reddish wash on the tail musculature and bold dark blotches on the tail fin, and lack an acuminate tip.

Green Treefrogs are similar in appearance to Barking Treefrogs (*D. gratiosa*), Pine Barrens Treefrogs (*D. andersonii*), and Squirrel Treefrog

(*D. squirella*). However, in Barking Treefrogs the dorsum has dark spots or blotches, the dorsal skin is bumpy, and the lateral light stripe has scalloped edges, becoming interrupted posteriorly. In Pine Barrens Treefrogs the sides of the body are dark maroon, lacking any evidence of wide light stripes. In the green morph of Squirrel Treefrog, the light lateral stripe has edges that shade gradually into the dorsal ground color, creating a diffuse edge to any light lateral stripe.

ALABAMA DISTRIBUTION Green Treefrog occurrence records are found throughout the state. Mount (1975) noted their absence in the Tennessee Valley and Highland Rim. However, recent records document their occupancy of the entire Tennessee Valley, suggesting a northward range expansion.

HABITS Green Treefrogs are common in a variety of permanently aquatic situations, including lakes, ponds, swamps, and some streams, where the species spends the breeding and nonbreeding seasons. They are also found in shallower wetlands such as marshes, cypress savannas, and cypress-gum swamps (Liner et al. 2008). Additionally, the species is common on barrier islands, suggesting that it is tolerant of brackish water. Sites with abundant emergent vegetation, especially cattails, are favored. During the day these frogs sit quietly on green leaves or stems, with their legs folded beneath them. This position renders them virtually invisible. Homes placed near wetlands occupied by Green Treefrogs may have this species visit outdoor lights at night to feed on insects. During winter these frogs seek refuge in rock

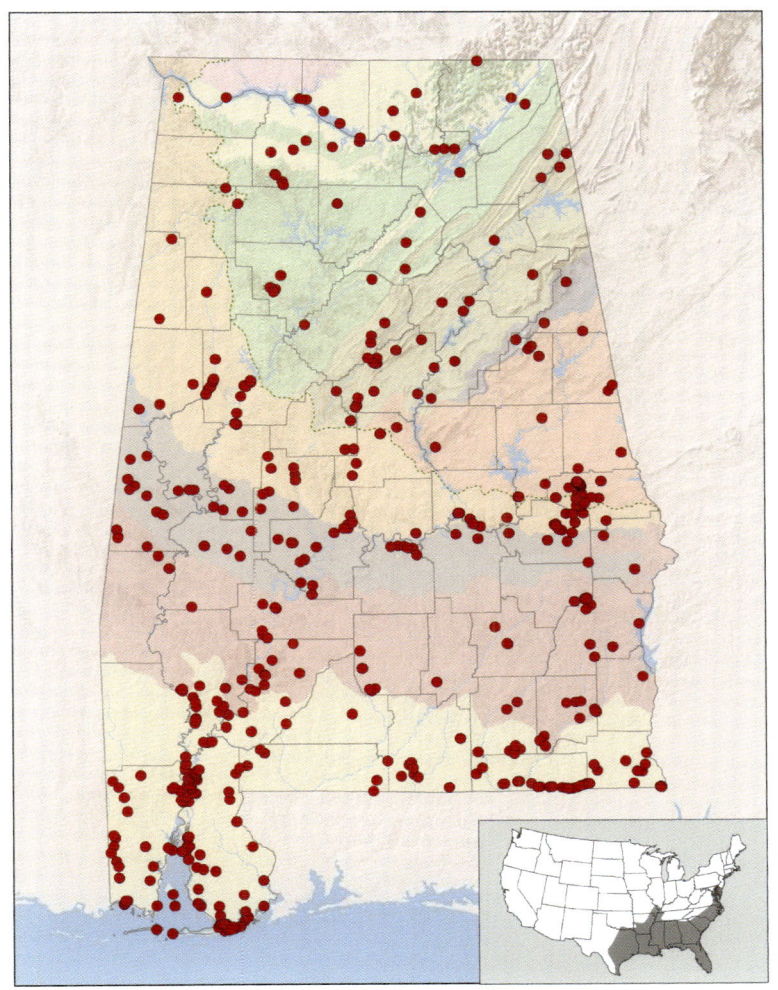

Distribution of Green Treefrog (*Dryophytes cinerea*). Solid dots indicate localities of specimens or photographs examined by the authors or ADCNR/Natural Heritage Program occurrence records believed to be valid. Inset map depicts approximate range in the United States.

piles, trash piles, facing boards of homes, under logs, and under the bark of dead trees.

The advertisement call of males, a "quunk" or "quonk," is issued at a rate of 30–60 notes per minute, usually from a perch 1–6 ft (0.3–2.0 m) above water. In Alabama, calling usually begins in April and extends into August. Rainy weather is not necessary to initiate vocalizations in this species. Green Treefrogs tend to call persistently throughout the breeding season (Steen et al. 2013), with calling activity peaking from dusk through midnight (Bridges and Dorcas 2000). Neighboring males may time their calls so that a duet of non-overlapping calls is made, but this typically is done in a fashion that does not make one male more attractive than the other (Höbel 2011). Non-calling satellite males frequently are found near calling males and have been observed to amplex gravid females that approach the calling male (Perrill et al. 1978). Satellite males are in poorer physical condition than calling males and have higher levels of stress hormones (Leary and Harris 2012). Once in amplexus, females choose a nest site in water where eggs are released and fertilized externally by the male. Up to 2,600 bicolored brown-and-cream eggs are produced by a female during a reproductive season (Gunzburger 2006), and these are deposited in packets of 10–20 eggs that float on the water's surface (Brown 1956). Tadpoles are known to be unpalatable to typical tadpole predators, allowing this species to reproduce in relatively permanent wetlands (Adams et al. 2011).

Diet of adults likely includes insects and other arthropods. Tadpoles eat periphyton and phytoplankton that are scraped from aquatic vegetation. Tadpoles may also scavenge nutrients from dead tadpoles.

Green Treefrogs are known to hybridize with Pine Barrens Treefrogs and Barking Treefrogs (Anderson and Moler 1986; Höbel and Gerhardt 2003). In places where Green Treefrogs and Barking Treefrogs attempt to breed in the same wetland, Green Treefrog males call from higher perches than they do at sites where Barking Treefrogs are not present. Females collected from sites where males of both species call show a stronger preference for calls of their own species than do females from sites occupied only by Green Treefrogs (Höbel and Gerhardt 2003). This character displacement prevents mismatings in areas where native vegetation is retained. However, in areas where pond margins are managed by removing riparian vegetation, Green Treefrog males call from the water where female Barking Treefrogs frequently select them as mates when only Green Treefrogs are calling (Oldham and Gerhardt 1975). Male Barking Treefrogs call from the water as well, but female Green Treefrogs tend not to mate with them when only Barking Treefrogs are calling. The mismatings that occur between male Green Treefrogs and female Barking Treefrogs produce viable first-generation hybrids (Lamb and Avise 1986).

Tadpoles that are hybrids of Green Treefrogs and Barking Treefrogs are palatable to fish and, therefore, suffer increased predation via fish predators in permanent ponds; thus, the adults that return to these breeding sites tend to be pure Green Treefrogs (Gunzburger 2005), a process that may eventually eliminate hybridization at that site.

CONSERVATION AND MANAGEMENT Green Treefrogs are abundant and have apparently viable populations in areas such as cities, county parks, and agricultural fields where habitat has been modified heavily by human activities (Babbitt and Tanner 2000, Guzy et al. 2012, Pham et al. 2007). For this reason, this species receives no regulatory protection in Alabama. Additionally, the species appears to require no special management tools to maintain it in Alabama's landscape. However, these frogs have experienced mortality events associated with high prevalence of ranavirus (Miller et al. 2011). Green Treefrogs can be infected by chytrid fungus (*Batrachochytrium dendrobatidis*) in lab settings but experience no deleterious effects to growth or survival (Brannelly et al. 2012). Chytrid has not been documented from free-ranging frogs.

TAXONOMY No subspecies of Green Treefrogs are recognized, but there is geographic variation in hind limb length such that specimens from the Appalachian Plateaus and Ridge and Valley regions of Alabama have shorter relative hind legs than specimens from the Piedmont and Coastal Plain (Aresco 1996). This species is the sister taxon to Barking Treefrogs (Faivovich et al. 2005).

Adult Barking Treefrog (*Dryophytes gratiosa*), Marengo County, AL.

Barking Treefrog
Dryophytes gratiosa (Le Conte, 1857 "1856")

DESCRIPTION Barking Treefrogs are large, plump hylids, attaining a maximum snout–vent length of around 2.8 in (70 mm). The toes end in large adhesive discs, and the dorsal surface is noticeably granular. The dorsal color is variable with most specimens being green or greenish but with some individuals being brown. On this ground color are a series of rounded dark spots, but these may be only faintly discernible, depending on dispersion of the pigments creating the ground color. On each side is a ragged light stripe or longitudinal series of irregular light markings and occasionally some irregular purplish or maroon markings. Tadpoles of this species are unmistakable because of their unusually tall tail fins. This feature makes the body shape nearly as tall as it is long. The body color of each tadpole is light silver with a dark smoky gray spot in the center of the tail musculature in young individuals that gradually disappears in older tadpoles.

As adults, Barking Treefrogs are most like Green Treefrogs (*D. cinerea*), but that species has smooth skin, a bold lateral stripe, and lacks dark dorsal spots. As tadpoles, Barking Treefrogs are most like Ornate Chorus Frogs (*Pseudacris ornata*) because both have a decidedly tall tail fin. However, Ornate Chorus Frogs lack a smoky gray spot on the tail musculature that is present on Barking Treefrog tadpoles when both are likely to be present together.

Adult Barking Treefrog (*Dryophytes gratiosa*), Marengo County, AL.

ALABAMA DISTRIBUTION This species is widely distributed across the Lower Coastal Plain, occurs in the Black Belt as far west as Dallas County, occupies the Fall Line Hills as far west as Tuscaloosa County, penetrates the Ridge and Valley as far north as Calhoun County, occurs sporadically throughout the Appalachian Plateau south of the Tennessee River, and occurs along the Tennessee River in Lauderdale County. Records for Barking Treefrogs are lacking from much of western Alabama, and the species is found only in peripheral localities within the Highland Rim, Piedmont, and Red Hills formations.

HABITS Shallow, semi-permanent ponds with at least some open water are the habitats most suitable for breeding of Barking Treefrogs (Liner et al. 2008). However, permanent ponds and lakes are used occasionally, if fish are lacking (Mount 1975). Adults migrate from surrounding forested areas to breeding sites, open shallow wetlands, at night during and immediately after rain showers (Todd and Winne 2006). We have heard males of this species calling as early as February 23 in south Alabama, and calling continues through August. During the calling season males vocalize for periods of one to four hours in water at the edge of seasonal wetlands (Murphy 1994). The call, a hollow "boonk" or "moonk," repeated every second or so, occurs sporadically throughout the breeding season, although calling activity is most intense from dusk to midnight (Bridges and Dorcas 2000). These calls are energetically costly to produce, and because of this, males lose

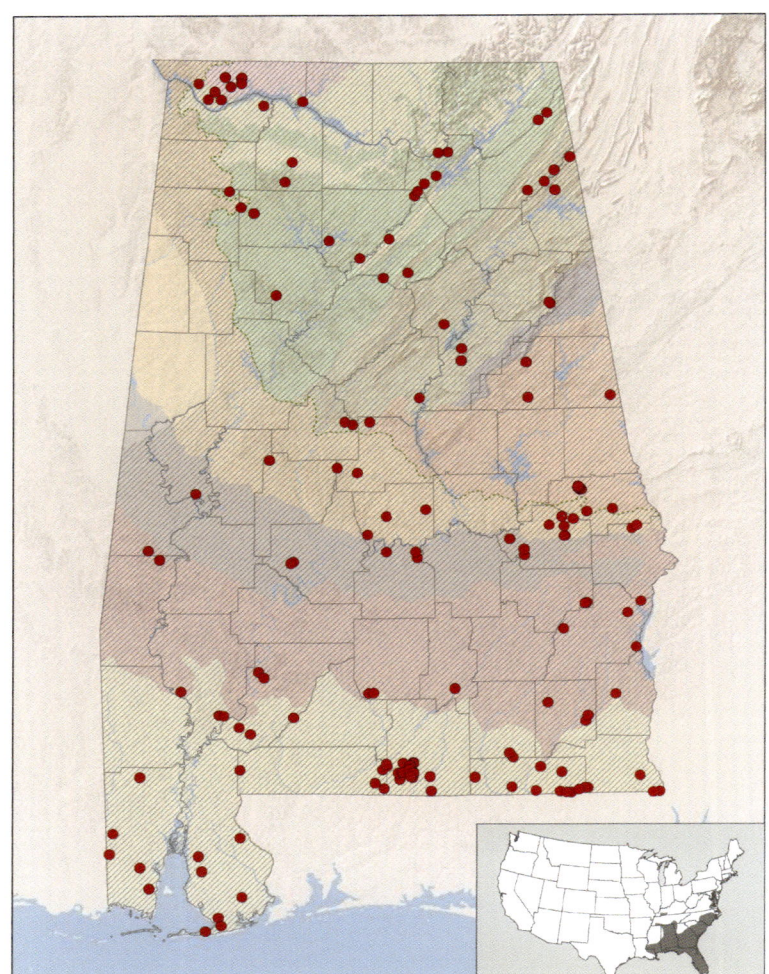

Distribution of Barking Treefrog (*Dryophytes gratiosa*). The presumed range of the species in Alabama is indicated by hatching. Solid dots indicate localities of specimens or photographs examined by the authors or ADCNR/Natural Heritage Program occurrence records believed to be valid. Inset map depicts approximate range in the United States.

weight while in the calling chorus and can maintain a position in it for only two to three nights (Murphy 1994). Females arrive later than males, prefer males with low dominant frequencies (deep voices), longer call durations, and faster call repetition rates (Burke and Murphy 2007). Males with these call features tend to be large (Poole and Murphy 2007), and likely arrive at the breeding site early each night (Murphy 1992). When a male is selected, the female nudges him with her snout, he ceases calling and grasps the female in pectoral amplexus. Satellite males are rare in this species as is aggression among males

(Murphy and Gerhardt 1996). A female selects nest sites that are underwater where she lays up to 4,000 eggs that are fertilized externally by the male. The eggs are laid singly and sink to the bottom of the water column. After breeding, adults move to forested areas adjacent to the breeding sites where they feed in trees and shrubs at night and burrow in the moist soil of grasses during the day. The species remains in burrows that it has created or enters mouse burrows, Gopher Tortoise (*Gopherus polyphemus*) burrows, or stump holes during winter, where they remain until warm weather in spring.

In places where Barking Treefrogs and Green Treefrogs attempt to breed in the same wetland and the wetland retains native shrubs around its edge, hybridization is rare because Green Treefrog males call from the shrubs and Barking Treefrog males call from the water. However, in areas where pond margins are managed by removing riparian vegetation, males of both species call from the water, and hybridization in which Barking Treefrog females select male Green Treefrogs may be frequent (Lamb and Avise 1986). Such mismatings occur on nights when Green Treefrogs are the only species calling (Oldham and Gerhardt 1975).

Diet of adults likely includes insects and other arthropods. Tadpoles eat periphyton and phytoplankton that are scraped from aquatic vegetation. Tadpoles may also scavenge nutrients from dead tadpoles.

Tadpoles of Barking Treefrogs are palatable to fish, and because of this the species tends to breed in temporary pools of water where fish are absent and many other hylid competitors are present. Tadpoles resulting from hybridization between Green Treefrogs and Barking Treefrogs (*D. cinerea* x *gratiosa*) are palatable to fish and, therefore, suffer increased predation via fish in permanent ponds. Thus, hybrid adults tend to be eliminated from permanent wetlands but accumulate in temporary wetlands creating a stronger effect of hybridization on Barking Treefrog than Green Treefrog populations (Gunzburger 2005).

CONSERVATION AND MANAGEMENT Barking Treefrogs are abundant, but sporadically distributed. Their general abundance causes them to receive no regulatory protection in Alabama. Because the species requires fish-free wetlands, it tends not to have viable populations in areas, such as city and county parks, where habitat has been modified heavily by human activities. In fact, the presence of Barking

Treefrogs is an indicator of intact natural wetlands (Guzy et al. 2012). Barking Treefrogs may occur in agricultural areas where they breed in borrow pits, roadside ditches, and fish-free farm ponds, but are detected more frequently in forested areas that are managed as open stands with a grass-dominated understory maintained with frequent fire (Babbitt and Turner 2000, Babbitt et al. 2006, Delis et al. 1996). The construction of farm ponds and the ecological alterations associated with disturbance on natural wetlands probably are causing the breakdown of the biological isolating mechanisms that served to keep Barking Treefrogs from hybridizing with Green Treefrogs in their unaltered natural habitats. Maintenance of native shrubs at the periphery of seasonal wetlands is vital to maintaining pure stocks of Barking Treefrogs.

TAXONOMY This species is sister to Green Treefrogs and is a member of the *cinerea* group (Faivovich et al. 2005). No subspecies are recognized for this species.

Glossary

ALLOPATRY: Located in or originating in different geographical areas.

AMPLEXUS: The behavior exhibited by male frogs during mating, involving a male grasping a female.

CAUDAL: Of or toward the tail.

CHERT: A type of sedimentary rock with silica forming microcrystals, cryptocrystals, or microfibers.

CHYTRIDIOMYCOSIS: Disease caused by chytrid fungus (*Batrachochytrium dendrobatidis*).

CLADE: a group containing an ancestor and all descendants of that ancestor

CLEAR-CUT: A method of forest management that involves removing all trees and replanting with seedlings.

CLINE: A region across which organisms or geology changes gradually.

CORNIFIED: Skin that is thickened, hardened, and frequently darkened.

CRYPTIC: Hidden from view.

DORSOLATERAL: Anatomical region above the middle of the side of the body.

DORSOLATERAL FOLD: A fold of skin running the length of each side of the body of some anurans from the tympanum to the pelvic girdle.

ENDRIN: Chlorinated hydrocarbon insecticide.

FLATWOODS: A habitat of the Coastal Plain dominated by pines and grasses and flooded seasonally by heavy rains.

FOSSORIAL: Living underground; burrowing.

FROGLET: A recently transformed anuran.

GRANULAR: Bumps on the skin that are about the size of grains of sand.

GUTTURAL: Of the throat.

HERPETOFAUNA: The assemblage of amphibian and reptile species at a particular locality or region.

HETEROSPECIFIC: An individual of a different species.

HYDROPERIOD: A measure of time during which a wetland contains water.

INTERORBITAL: Anatomical region on the top of the head between the eyes.

LEK: A mating system in which males aggregate to attract females who are attracted to individual males only by the information presented by the male vocalization and/or behavior.

LEPTODACTYLIFORM: Any member of the frog families Allophrynidae, Calyptocephalellidae, Centrolenidae, Ceratophridae, Cycloramphidae, Hylodidae, Leiuperidae, and Leptodactylidae.

LINNAEAN HIERARCHY: The system of taxonomic groups invented by Carolus Linnaeus and forming the basis of modern classifications.

METAMORPHS: Newly transformed anurans.

METAPOPULATION: A group of geographically contiguous populations that share individuals via migration.

MIS-MATING: Mating with the wrong species.

MONOPHYLETIC: A group in which all members are more closely related to other group members than to any individual outside the group.

OVERSTORY: The portion of the vegetation that forms the canopy.

PAPILLATE: Covered with numerous finger-like projections.

PAROTOID GLAND: A skin gland of some anurans located behind the eye and producing toxic secretions.

PERIPHYTON: The group of microscopic organisms that cover aquatic plants.

PHYLOGEOGRAPHY: The study of patterns of gene flow among populations of the same species across broad geographic scales.

PHYSIOGRAPHY: Non-random patterns of soils and vegetation across broad geographic scales.

PHYTOPLANKTON: Single-celled organisms capable of photosynthesis and suspended within a water column.

PLANKTIVOROUS: Consuming plankton.

POSTANAL: Anatomical region posterior to the cloacal opening of anurans.

POSTORBITAL: Anatomical region immediately behind an eye.

PREMETAMORPHIC: Developmental time period before metamorphosis from the tadpole stage to the adult body form.

PUSTULATE: Containing rounded protuberances.

SUBGENUS: A group of closely related species within a genus.

SUBORDER: A group of closely related families within an order.

SUBSPECIFIC: A taxonomic category designating a geographic variant within a species.

SUPERFAMILY: A group of closely related families.

SUPRAORBITAL: Anatomical region immediately dorsal or medial to an eye.

SYMPATRY: Located in or originating in the same geographical area.

TETRAPOD: Any member of the group Tetrapoda, the land vertebrates.

TIBIAL: Associated with the portion of the hind limb of anurans containing the tibia.

TIBIO-TARSUS: A bone formed by the fusion of the tibia with a tarsal element.

TOXAPHENE: A chlorinated molecule used as an insecticide.

UROSTYLE: A unique bone of the pelvic girdle of anurans.

VAGILE: Wide-ranging, moving easily from place to place.

VENTER: The belly of an organism.

VERMICULATIONS: Wormlike markings.

WARTY: Bumps on the skin that are larger than the size of sand grains.

Photo Credits

Johnny Autery, pages 111, 167, 171, 179, 187, 206, 210–11

Mark Bailey, pages 11, 16–17, 18 (bottom), 24, 28, 30–31, 60, 106, 175

Mary Dansak, page vii

Brian Folt, page 53

Brad Glorioso, page 142

James C. Godwin, pages 64, 73, 80, 98, 102, 154

Sean Graham, pages 46, 68

Scott Gravette, pages 101, 205

Matt Greene, page 132

Craig Guyer, pages 12, 18 (top), 19–23, 25–26, 29, 32, 133

Aubrey M. Heupel, pages 161, 202

Pierson Hill, pages 128, 164

Brian D. Holt, ADCNR, pages 10, 93

Ritchie King, page 150

Kevin Messenger, page 176

Todd Pierson, page 139

Corey Raimond, page 195

Eric Soehren, pages 107, 121

Sierra Stiles, page 13

Kevin Stohlgren, page 191

John A. Tupy, page 116

Kenneth Wray, pages 125, 192, 198

Works Cited

Adams, C. K., and D. Saenz. 2012. "Leaf litter of invasive Chinese tallow tree
(*Triadica sebifera*) negatively affects hatching success of an aquatic breed-
ing amphibian, the southern leopard frog (*Lithobates sphenocephalus*)."
Canadian Journal of Zoology 90:991–998.

Adams, C. K., D. Saenz, and R. N. Conner. 2011. "Palatability of twelve
species of anuran larvae in eastern Texas." *American Midland Naturalist*
166:211–223.

Alabama Department of Conservation and Natural Resources, Division of
Wildlife and Freshwater Fisheries. *Alabama's Wildlife Action Plan, 2015–
2025.* 2015, amended 2016. www.outdooralabama.com.

Alix, D. M., C. J. Anderson, J. B. Grand, and C. Guyer. 2014a. "Evaluating
the effects of land use on headwater wetland amphibian assemblages in
coastal Alabama." *Wetlands* 34:917–926.

Alix, D. M., C. Guyer, and C. J. Anderson. 2014b. "Expansion of the range of
the introduced Greenhouse Frog, *Eleutherodactylus planirostris*, in coastal
Alabama." *Southeastern Naturalist* 13:N59–N62.

Anderson, A., and P. E. Moler. 1986. "Natural hybrids of the Pine Barrens
Treefrog, *Hyla andersonii* with *H. cinerea* and *H. femoralis* (Anura, Hylidae): Morphological and chromosomal evidence." *Copeia* 1986:70–76.

Aresco, M. J. 1996. "Geographic variation in the morphology and lateral
stripe of the Green Treefrog (*Hyla cinerea*) in the southeastern United
States." *American Midland Naturalist* 135:293–298.

Ashton, R. E., and P. Ashton. 1988. *Handbook of Reptiles and Amphibians of
Florida.* Part 3, *The Amphibians.* Miami, FL: Windward.

Austin, J. D., J. A. Dávila, S. C. Lougheed, and P. T. Boag. 2003. "Genetic
evidence for female-biased dispersal in the bullfrog, *Rana catesbeiana* (Ranidae)." *Molecular Ecology* 12:3165–3172.

Austin, J. D., S. C. Lougheed, L. Neidrauer, A. A. Chek, and P. T. Boag. 2002.
"Cryptic lineages in a small frog: The post-glacial history of the spring
peeper, *Pseudacris crucifer* (Anura: Hylidae)." *Molecular Phylogenetics and
Evolution* 25:316–329.

Austin, J. D., and K. R. Zamudio. 2008. "Incongruence in the pattern and
timing of intraspecific diversification in bronze frogs and bullfrogs (Ranidae)." *Molecular Phylogenetics and Evolution* 48:1041–1053.

Babbitt, K. J., M. J. Baber, and L. A. Brandt. 2006. "The effect of woodland proximity and wetland characteristics on larval anuran assemblages in an agricultural landscape." *Canadian Journal of Zoology* 84:510–519.

Babbitt, K. J., M. J. Baber, D. L. Childers, and D. Hocking. 2009. "Influence of agricultural upland habitat type on larval anuran assemblages in seasonally inundated wetlands." *Wetlands* 29:294–301.

Babbitt, K. J., and G. W. Tanner. 2000. "Use of temporary wetlands by anurans in a hydrologically modified landscape." *Wetlands* 20:313–322.

Bailey, M. A. 1991. "Migration of *Rana areolata sevosa* and associated winter-breeding amphibians at a temporary pond in the Lower Coastal Plain of Alabama." Master's thesis, Auburn University.

Baud, D. R., and M. L. Beck. 2005. "Interactive effects of UV-B and copper on spring peeper tadpoles (*Pseudacris crucifer*)." *Southeastern Naturalist* 4:15–22.

Baughman, B., and B. D. Todd. 2007. "Role of substrate cues in habitat selection by recently metamorphosed *Bufo terrestris* and *Scaphiopus holbrookii*." *Journal of Herpetology* 41:154–157.

Beasley, V. R., S. A. Faeh, B. Wikoff, J. Eisold, D. Nichols, R. Cole, A. M. Schotthoefer, C. Staehle, M. Greenwell, and L. W. Brown. 2005. "Risk factors and the decline of the northern cricket frog, *Acris crepitans*: Evidence for the involvement of herbicides, parasitism, and habitat modifications." In *The Status and Conservation of United States Amphibians*, ed. M. Lannoo, 75–87. Iowa City: University of Iowa Press.

Bee, M. A., C. E. Kozich, K. J. Blackwell, and H. C. Gerhardt. 2001. "Individual variation in advertisement calls of territorial male green frogs, *Rana clamitans*: Implications for individual recognition." *Ethology* 107:65–84.

Berven, K. A. 1990a. "Dispersal in the wood frog (*Rana sylvatica*): Implications for genetic population structure." *Evolution* 44:2047–2056.

Berven, K. A. 1990b. "Factors affecting population fluctuations in larval and adult stages of the wood frog (*Rana sylvatica*)." *Ecology* 71:1599–1608.

Berven, K. A. 1995. "Population regulation in the wood frog, *Rana sylvatica*, from three diverse geographic localities." *Australian Journal of Ecology* 20:385–392.

Binckley, C. A., and W. J. Resetarits Jr. 2002. "Reproductive decisions under threat of predation: Squirrel treefrog (*Hyla squirella*) responses to banded sunfish (*Ennacanthus obesus*)." *Oecologia* 130:157–161.

Binckley, C. A., and W. J. Resetarits Jr. 2003. "Functional equivalence of non-lethal effects: Generalized fish avoidance determines distribution of gray treefrog, *Hyla chrysoscelis*, larvae." *Oikos* 102:623–629.

Birkhead, R. D., J. P. McGuire, R. Conley, and C. K. Ward. 2017. "*Incilius nebulifer* (Gulf Coast Toad)." *Herpetological Review* 48:120.

Birx-Raybuck, D. A., S. J. Price, and M. E. Dorcas. 2009. "Pond age and riparian zone proximity influence anuran occupancy of urban retention ponds." *Urban Ecosystems* 13:181–190.

Blaustein, A. R., L. K. Bleden, D. H. Olson, D. M. Green, T. L. Root, and J. M. Kiesecker. 2001. "Amphibian breeding and climate change." *Conservation Biology* 15:1804–1809.

Blihovde, W. B. 2006. "Terrestrial movements and upland habitat use of gopher frogs in central Florida." *Southeastern Naturalist* 5:265–276.

Blouin, M. S. 1989. "Life history correlates of a color polymorphism in the ornate chorus frog, *Pseudacris ornata*." *Copeia* 1989:319–325.

Boone, M. D., S. A. Hammond, N. Veldhoen, M. Youngquist, and C. C. Helbing. 2013. "Specific time of exposure during tadpole development influences biological effects of the insecticide carbaryl in green frogs (*Lithobates clamitans*)." *Aquatic Toxicology* 130–131:139–148.

Boone, M. D., and S. M. James. 2003. "Interactions of an insecticide, herbicide, and natural stressors in amphibian community mesocosms." *Ecological Applications* 13:829–841.

Brannelly, L. A. 2014. "Reduced itraconazole concentration and durations are successful in treating *Batrachochytrium dendrobatidis* infections in amphibians." *Journal of Visual Experiments* (85):e51166.

Brannelly, L. A., M. W. H. Chatfield, and C. L. Richards-Zawacki. 2012. "Field and laboratory studies of the susceptibility of the green treefrog (*Hyla cinerea*) to *Batrachochytrium dendrobatidis* infection." *PLoS One* 7(6):e38473.

Breden, F., and C. H. Kelly. 1982. "The effect of conspecific interactions on metamorphosis in *Bufo americanus*." *Ecology* 63:1682–1689.

Bridges, A. S., and M. E. Dorcas. 2000. "Temporal variation in anuran calling behavior: Implications for surveys and monitoring programs." *Copeia* 2000:587–592.

Bridges, C. M. 2000. "Long-term effects of pesticide exposure at various life stages of the southern leopard frog (*Rana sphenocephala*)." *Archives of Environmental Contamination and Toxicology* 39:91–96.

Brodie, E. D., Jr., and D. R. Formanowicz Jr. 1987. "Antipredator mechanisms of larval amphibians: Protection of palatable individuals." *Herpetologica* 43:369–375.

Brodie, E. D., Jr., D. R. Formanowicz Jr., and E. D. Brodie III. 1978. "The development of noxiousness of *Bufo americanus* tadpoles to aquatic insect predators." *Herpetologica* 34:302–306.

Brown, J. S. 1956. "The frogs and toads of Alabama." PhD diss., University of Alabama.

Brown, L. E., and D. B. Means. 1984. "Fossorial behavior and ecology of the chorus frog *Pseudacris ornata*." *Amphibia-Reptilia* 5:261–273.

Burke, E. J., and C. G. Murphy. 2007. "How female barking treefrogs, *Hyla gratiosa*, use multiple call characteristics to select a mate." *Animal Behaviour* 74:1463–1472.

Burmeister, S. S., J. Konieczka, and W. Wilczynski. 1999. "Agonistic encounters in a cricket frog (*Acris crepitans*) chorus: Behavioral outcomes vary with local competition and within the breeding season." *Ethology* 105:335–347.

Burmeister, S. S., A. G. Ophir, M. J. Ryan, and W. Wilczynski. 2002. "Information transfer during cricket frog contests." *Animal Behaviour* 64:715–725.

Caldwell, J. P. 1982. "Disruptive selection: A tail color polymorphism in *Acris* tadpoles in response to differential predation." *Canadian Journal of Zoology* 60:2818–2827.

Carey, S. D. 1982. "Geographic distribution: *Eleutherodactylus planirostris.*" *Herpetological Review* 13:130.

Chivers, J. M. 2016. "Combining morphological, acoustic, and genetic techniques to better understand hybridization of the most abundant toad in Alabama, *Anaxyrus fowleri.*" Master's thesis, Auburn University.

Christman, S. P. 1970. "*Hyla andersonii* in Florida." *Quarterly Journal of the Florida Academy of Sciences* 33:80.

Churchill, T. A., and K. B. Storey. 1996. "Organ protection and cryoprotectant synthesis during freezing in spring peepers *Pseudacris crucifer.*" *Copeia* 1996:517–525.

Conaway, C. H., and D. E. Metter. 1967. "Skin glands associated with breeding in *Microhyla carolinensis.*" *Copeia* 1967:672–673.

Cosentino, B. J., D. M. Marsh, K. S. Jones, J. J. Apodaca, C. Bates, J. Beach, K. H. Beard, K. Becklin, J. M. Bell, C. Crockett, G. Fawson, J. Fjelsted, E. A. Forys, K. S. Genet, M. Grover, J. Holmes, K. Indeck, N. E. Karraker, E. S. Kilpatrick, T. A. Langen, S. G. Mugel, A. Molina, J. R. Vonesh, R. J. Weaver, and A. Willey. 2014. "Citizen science reveals widespread negative effects of roads on amphibian distributions." *Biological Conservation* 180:31–38.

Costanzo, J. P., M. F. Wright, and R. E. Lee Jr. 1992. "Freeze tolerance as an overwintering adaptation in Cope's Grey Treefrog (*Hyla chrysoscelis*)." *Copeia* 1992:565–569.

Cotton, T. B., M. A. Kwiatkowski, D. Saenz, and M. Collyer. 2012. "Effects of an invasive plant, Chinese tallow (*Triadica sebifera*), on development and survival of anuran larvae." *Journal of Herpetology* 46:186–193.

Crossland, M. R., G. P. Brown, M. Anstis, C. M. Shilton, and R. Shine. 2008. "Mass mortality of native anuran tadpoles in tropical Australia due to the invasive cane toad (*Bufo marinus*)." *Biological Conservation* 141:2387–2394.

Crother, B. I. (Committee Chair). 2012. *Scientific and Standard English Names of Amphibians and Reptiles of North America North of Mexico, with Comments regarding Confidence in Our Understanding.* 7th ed. SSAR Herpetological Circular 37. Shoreview, MN: SSAR.

Crother, B. I. (Committee Chair). 2017. *Scientific and Standard English*

Common Names of Amphibians and Reptiles of North America North of Mexico, with Comments regarding Confidence in Our Understanding. 8th ed. SSAR Herpetological Circular 43. Shoreview, MN: SSAR.

Daszak, P., D. E. Scott, A. M. Kilpatrick, C. Faggioni, J. W. Gibbons, and D. Porter. 2005. "Amphibian population declines at Savannah River Site are linked to climate, not chytridiomycosis." *Ecology* 86:3232–3237.

Davis, J. R., D. T. Eastlack, A. J. Kouba, and C. K. Vance. 2012. "*Batrachochytrium dendrobatidis* detected in Fowler's toad (*Anaxyrus fowleri*) populations in Memphis, Tennessee, USA." *Herpetological Review* 43:150–159.

Degner, J. F., D. M. Silva, T. D. Hether, J. M. Daza, and E. A. Hoffman. 2010. "Fat frogs, mobile genes: Unexpected phylogeographic patterns for the ornate chorus frog (*Pseudacris ornata*)." *Molecular Ecology* 19:2501–2515.

Delis, P. R., H. R. Mushinsky, and E. D. McCoy. 1996. "Decline of some west-central Florida anuran populations in response to habitat degradation." *Biodiversity and Conservation* 5:1579–1595.

De Queiroz, K., and J. Gauthier. 1992. "Phylogenetic taxonomy." *Annual Review of Ecology and Systematics* 23:449–480.

De Sá, R. O., J. W. Streicher, R. Sekonyela, M. C. Forlani, S. P. Loader, E. Greenbaum, S. Richards, and C. F. B. Haddad. 2012. "Molecular phylogeny of microhylid frogs (Anura: Microhylidae) with emphasis on relationships among New World genera." *BMC Evolutionary Biology* 12:241.

Deyrup, M., L. Deyrup, and J. Carrel. 2013. "Ant species in the diet of a Florida population of Eastern Narrow-mouthed Toads, *Gastrophryne carolinensis*." *Southeastern Naturalist* 12:367–378.

Dodd, C. K. 1994. "The effects of drought on population structure, activity, and orientation of toads (*Bufo quercicus* and *B. terrestris*) at a temporary pond." *Ethology Ecology & Evolution* 6:331–349.

Dodd, C. K., and B. S. Cade. 1998. "Movement patterns and the conservation of amphibians breeding in small, temporary wetlands." *Conservation Biology* 12:331–339.

Duellman, W. E., A. B. Marion, and S. B. Hedges. 2016. "Phylogenetics, classification, and biogeography of the treefrogs (Amphibia: Anura: Arboranae)." *Zootaxa* 4104:1–109.

Duellman, W. E., and L. R. Trueb. 1986. *Biology of Amphibians.* New York: McGraw-Hill.

Dundee, H. A., D. A. Rossman, and E. C. Beckham. 1996. *Amphibians and Reptiles of Louisiana.* Baton Rouge: Louisiana State University Press.

Edge, C. B., M. E. Gahl, B. D. Pauli, D. G. Thompson, and J. E. Houlahan. 2011. "Exposure of juvenile green frogs (*Lithobates clamitans*) in littoral enclosures to a glyphonsate-based herbicide." *Ecotoxicology and Environmental Safety* 74:1363–1369.

Edwards, T. M., K. A. McCoy, T. Barbeau, M. W. McCoy, J. M. Thro, and L. J.

Guillette Jr. 2006. "Environmental context determines nitrate toxicity in southern toad (*Bufo terrestris*) tadpoles." *Aquatic Toxicology* 78:50–58.

Engbrecht, N. J., and M. J. Lannoo. 2010. "A review of the status and distribution of Crawfish Frogs (*Lithobates areolatus*) in Indiana." *Proceedings of the Indiana Academy of Sciences* 119:64–73.

Engbrecht, N. J., M. J. Lannoo, P. J. Williams, J. R. Robb, T. A. Gerardot, D. R. Karns, and M. J. Lodato. 2013. "Is there hope for the Hoosier frog? An update on the status of Crawfish Frogs (*Lithobates areolatus*) in Indiana, with recommendations for their conservation." *Proceedings of the Indiana Academy of Sciences* 121:147–157.

Erdmann, J. A. 2017. "The function of toe movement in feeding by the Gulf Coast toad (*Incilius nebulifer*)." Master's thesis, Southeastern Louisiana University.

Faivovich, J., C. F. B. Haddad, P. C. A. Garcia, D. R. Frost, J. A. Campbell, and W. C. Wheeler. 2005. "Systematic review of the frog family Hylidae, with special reference to Hylinae: A phylogenetic analysis and taxonomic revision." *Bulletin of the American Museum of Natural History* 294:1–240.

Ferguson, D. E., H. F. Landreth, and J. P. McKeown. 1967. "Sun compass orientation of the northern cricket frog, *Acris crepitans*." *Animal Behaviour* 15:45–53.

Foley, D. H., III. 1994. "Short-term response of herpetofauna to timber harvesting in conjunction with streamside management zones in seasonally-flooded bottomland hardwood forests of southeast Texas." Master's thesis, Texas A&M University.

Fontenot, B. E., R. Makowsky, and P. T. Chippendale. 2011. "Nuclear-mitochondrial discordance and gene flow in a recent radiation of toads." *Molecular Phylogenetics and Evolution* 59:66–80.

Forester, D. C., J. W. Snodgrass, K. Marsalek, and Z. Lanham. 2006. "Post-breeding dispersal and summer home range of female American toads (*Bufo americanus*)." *Northeastern Naturalist* 13:59–72.

Franz, R., C. K. Dodd, and C. Jones. 1988. "Life history notes: *Rana areolata aesopus* (Florida gopher frog). Movement." *Herpetological Review* 19:82.

Freidenfelds, N. A., J. L. Purrenhage, and K. J. Babbitt. 2011. "The effects of clearcuts and forest buffer size on post-breeding emigration of adult wood frogs (*Lithobates sylvaticus*)." *Forest Ecology and Management* 261:2115–2122.

Frost, D. R. 2017. "Amphibian species of the world: An online reference." Version 6.0 (accessed Nov. 28, 2017). New York: American Museum of Natural History. http://research.amnh.org/herpetology/amphibia/index.html.

Frost, D. R., T. Grant, J. Faivovich, R. H. Bain, A. Haas, C. F. B. Haddad, R. O. de Sá, A. Channing, M. Wilkinson, S. C. Donnellan, C. J. Raxworthy, J. A. Campbell, B. L. Blotto, P. Moler, R. C. Drewes, R. A. Nussbaum, J. D.

Lynch, D. M. Green, and W. C. Wheeler. 2006. "The amphibian tree of life." *Bulletin of the American Museum of Natural History* No. 297:1–370.

Frost, D. R., and D. M. Hillis. 1990. "Species in concept and practice: Herpetological applications." *Herpetologica* 46:86–104.

Frost, D. R., E. M. Lemmon, R. McDiarmid, and J. R. Mendelson III. 2017. "Anura—Frogs and Toads." In *Scientific and Standard English Names of Amphibians and Reptiles of North America North of Mexico, with Comments regarding Confidence in Our Understanding*, 8th ed., B. I. Crother (Committee Chair), 6–24. SSAR Herpetological Circular 43. Shoreview, MN: SSAR.

Gahl, M. K., J. E. Longcore, and J. E. Houlahan. 2011. "Varying responses of northeastern North American amphibians to the chytrid pathogen *Batrachochytrium dendrobatidis*." *Conservation Biology* 26:135–141.

Gamble, T., P. B. Berendzen, H. B. Shaffer, D. E. Starkey, and A. M. Simons. 2008. "Species limits and phylogeography of North American cricket frogs (*Acris*: Hylidae)." *Molecular Phylogenetics and Evolution* 48:112–125.

Garton, J. S., and H. R. Mushinsky. 1979. "Integumentary toxicity and unpalatability as an antipredator mechanism in the narrow-mouthed toad, *Gastrophryne carolinensis*." *Canadian Journal of Zoology* 57:1965–1973.

Gerhardt, H. C. 1974. "Behavioral isolation of the tree frogs, *Hyla cinerea* and *Hyla andersonii*." *American Midland Naturalist* 91:424–433.

Gerhardt, H. C., M. L. Dyson, and S. D. Tanner. 1996. "Dynamic properties of the advertisement calls of gray tree frogs: Patterns of variability and female choice." *Behavioral Ecology* 7:7–18.

Gibbs, J. P. 1998. "Distribution of woodland amphibians along a forest fragmentation gradient." *Landscape Ecology* 13:263–268.

Godwin, J. C. 1985. "Extrinsic factors influencing the breeding of the pine barrens treefrog, *Hyla andersonii*." Master's thesis, Auburn University.

Goldberg, S. R. 2017. "Reproduction in Gulf Coast toads, *Incilius nebulifer* (Anura: Bufonidae) from Texas." *Bulletin of the Chicago Herpetological Society* 52:159–161.

Green, D. M. 2013. "Sex ratio and breeding population size in Fowler's toad, *Anaxyrus* (=*Bufo*) *fowleri*." *Copeia* 2013:647–652.

Greenberg, C. H. 2001. "Spatio-temporal dynamics of pond use and recruitment in Florida gopher frogs (*Rana capito aesopus*)." *Journal of Herpetology* 35:74–85.

Greenberg, C. H., and G. W. Tanner. 2004. "Breeding pond selection and movement patterns by eastern spadefoot toads (*Scaphiopus holbrookii*) in relation to weather and edaphic conditions." *Journal of Herpetology* 38:569–577.

Greenberg, C. H., and G. W. Tanner. 2005a. "Spatial and temporal ecology of eastern spadefoot toads on a Florida landscape." *Herpetologica* 61:20–28.

Greenberg, C. H., and G. W. Tanner. 2005b. "Spatial and temporal ecology of oak toads (*Bufo quercicus*) in a Florida landscape." *Herpetologica* 61:422–434.

Greenberg, D. A., and D. M. Green. 2013. "Effects of an invasive plant on population dynamics in toads." *Conservation Biology* 27:1049–1057.

Greenlees, M. J., G. P. Brown, J. K. Webb, B. L. Phillips, and R. Shine. 2006. "Effects of an invasive anuran [the cane toad (*Bufo marinus*)] on the invertebrate fauna of a tropical Australian floodplain." *Animal Conservation* 9:431–438.

Greenspan, S., A. J. K. Calhoun, J. E. Longcore, and M. G. Levy. 2012. "Transmission of *Batrachochytrium dendrobatidis* to wood frogs (*Lithobates sylvaticus*) via a bullfrog (*L. catesbeianus*) vector." *Journal of Wildlife Diseases* 48:575–582.

Gregoire, D. R., and M. S. Gunzburger. 2008. "Effects of predatory fish on survival and behavior of larval gopher frogs (*Rana capito*) and southern leopard frogs (*Rana sphenocephala*)." *Journal of Herpetology* 42:97–103.

Griffith, G. E., J. M. Omernik, J. A. Comstock, S. Lawrence, G. Martin, A. Goddard, V. J. Hulcher, and T. Foster. 2001. "Ecoregions of Alabama and Georgia." (2-sided color poster with map, descriptive text, summary tables, and photographs). U.S. Geological Survey, Reston, VA.

Gunzburger, M. S. 2005. "Differential predation on tadpoles influences the potential effects of hybridization between *Hyla cinerea* and *Hyla gratiosa*." *Journal of Herpetology* 39:682–687.

Gunzburger, M. S. 2006. "Reproductive ecology of the green treefrog (*Hyla cinerea*) in northwestern Florida." *American Midland Naturalist* 155:321–328.

Guyer, C., M. A. Bailey, J. Holmes, J. A. Stiles, and S. H. Stiles. 2007. "Herpetofaunal response to longleaf pine ecosystem restoration, Conecuh National Forest, Alabama." Final report, State Wildlife Grants, Alabama Division of Wildlife and Freshwater Fisheries.

Guzy, J. C., E. D. McCoy, A. C. Deyle, S. M. Gonzalez, N. Halstead, and H. R. Mushinsky. 2012. "Urbanization interferes with the use of amphibians as indicators of ecological integrity of wetlands." *Journal of Applied Ecology* 49:941–952.

Hall, R. J., and D. Swineford. 2003. "Toxic effects of endrin and toxaphene on the southern leopard frog *Rana sphenocephala*." *Environmental Pollution Series A, Ecological and Biological* 23:63–65.

Halstead, N. T. 2007. "Long-term effects of prescribed fire on reptile and amphibian communities in Florida sandhill." Master's thesis, University of South Florida.

Hanlon, S. M., J. L. Kerby, and M. J. Parris. 2012. "Unlikely remedy:

Fungicide clears infection from pathogenic fungus in larval southern leopard frogs (*Lithobates sphenocephalus*)." *PLoS One* 7(8):e43573.

Hanselmann, R., A. Rodríguez, M. Lampo, L. Fajardo-Ramos, A. A. Aguirre, A. M. Kilpatrick, J. P. Rodríguez, and P. Daszak. 2004. "Presence of an emerging pathogen of amphibians in introduced bullfrogs *Rana catesbeiana* in Venezuela." *Biological Conservation* 120:115–119.

Hayes-Odum, L. A. 1990. "Observations on reproduction and embryonic development in *Syrrhophus cystignathoides campi* (Anura: Leptodactylidae)." *Southwestern Naturalist* 35:358–361.

Hedges, D. B., W. E. Duellman, and M. P. Heinicke. 2008. "New World direct-developing frogs (Anura: Terrarana): Molecular phylogeny, classification, biogeography, and conservation." *Zootaxa* 1737:1–182.

Heemeyer, J. L., and M. J. Lannoo. 2012. "Breeding migrations in crawfish frogs (*Lithobates areolatus*): Long-distance movements, burrow philopatry, and mortality in a near-threatened species." *Copeia* 2012:440–450.

Heemeyer, J. L., P. J. Williams, and M. J. Lannoo. 2012. "Obligate crayfish burrow use and core habitat requirements of Crawfish Frogs." *Journal of Wildlife Management* 76:1081–1091.

Heinen, J. T. 1993. "Aggregation of newly metamorphosed *Bufo americanus*: Tests of two hypotheses." *Canadian Journal of Zoology* 71:334–338.

Heinicke, M. P., W. E. Duellman, L. Trueb, D. B. Means, R. D. MacCulloch, and S. B. Hedges. 2009. "A new frog family (Anura: Terrarana) from South America and an expanded direct-developing clade revealed by molecular phylogeny." *Zootaxa* 2211:1–35.

Hellman, R. E. 1953. "A comparative study of the eggs and tadpoles of *Hyla phaeocrypta* and *Hyla versicolor* in Florida." *Publications of the Research Division of Ross Allen's Reptile Institute* 1:61–74.

Hersikorns, B. D., and J. E. G. Smits. 2011. "Compromised metamorphosis and thyroid hormone changes in wood frogs (*Lithobates sylvaticus*) raised on reclaimed wetlands on the Athabasca oil sands." *Environmental Pollution* 159:596–601.

Hether, T. D., and E. A. Hoffman. 2012. "Machine learning identifies specific habitats associated with genetic connectivity in *Hyla squirella*." *Journal of Evolutionary Biology* 25:1039–1052.

Higley, E., A. R. Tompsett, J. P. Giesy, M. Hecker, and S. Wiseman. 2013. "Effects of triphenyltin on growth and development of the wood frog (*Lithobates sylvaticus*)." *Aquatic Toxicology* 144–145:155–161.

Hillis, D. M. 1988. "Systematics of the *Rana pipiens* complex: Puzzle and paradigm." *Annual Review of Ecology and Systematics* 19:39–63.

Hillis, D. M., and T. P. Wilcox. 2005. "Phylogeny of the New World true frogs (*Rana*)." *Molecular Phylogenetics and Evolution* 34:299–314.

Hinkson, K. N., and S. W. Richter. 2016. "Temporal trends in genetic data and effective population size support efficacy of management practices in critically endangered dusky gopher frogs (*Lithobates sevosus*)." *Ecology and Evolution* 6:2667–2678.

Höbel, G. 2011. "Variation in signal timing behavior: Implications for male attractiveness and sexual selection." *Behavioral Ecology and Sociobiology* 65:1283–1294.

Höbel, G., and H. C. Gerhardt. 2003. "Reproductive character displacement in the acoustic communication system of green tree frogs (*Hyla cinerea*)." *Evolution* 57:894–904.

Höbel, G., and R. C. Kolodziej. 2013. "Wood frogs (*Lithobates sylvaticus*) use water surface waves in their reproductive behaviour." *Behaviour* 150:471–483.

Hoffman, A. S., J. L. Heemeyer, P. J. Williams, J. R. Robb, D. R. Karns, V. C. Kinney, N. J. Engbrecht, and M. J. Lannoo. 2010. "Strong site fidelity and a variety of imaging techniques reveal around-the-clock and extended activity patterns in crawfish frogs (*Lithobates areolatus*)." *Bioscience* 60:829–834.

Holcombe, T., T. J. Stohlgren, and C. Jarnevich. 2008. "Invasive species management and research using GIS." In *Managing Vertebrate Invasive Species: Proceedings of an International Symposium*, ed. G. W. Witmer, W. C. Pitt, and K. A. Fagerstone, 108–114. Fort Collins, CO: USDA/APHIS Wildlife Service, National Wildlife Research Center.

Holloway, A. K., D. C. Cannatella, H. C. Gerhardt, and D. M. Hillis. 2006. "Polyploids with different origins and ancestors form a single sexual polyploid species." *American Naturalist* 167:E88–E101.

Holt, B. D. 2015. "*Lithobates areolatus* (Crawfish Frog)." *Herpetological Review* 46:378–379.

Homan, R. N., J. R. Bartling, R. J. Stenger, and J. L. Brunner. 2013. "Detection of ranavirus in Ohio, USA." *Herpetological Review* 44:615–618.

Hopkins, W. A., S. E. DuRant, B. P. Staub, C. L. Rowe, and B. P. Jackson. 2006. "Reproduction, embryonic development, and maternal transfer of contaminants in the amphibian *Gastrophryne carolinensis*." *Environmental Health Perspectives* 114:661–666.

Hoverman, J. T., M. J. Gray, N. A. Haislip, and D. L. Miller. 2011. "Phylogeny, life history, and ecology contribute to differences in amphibian susceptibility to ranaviruses." *EcoHealth* 8:301–309.

Howard, R. D. 1988. "Reproductive success in two species of anurans." In *Reproductive Success*, ed. T. H. Clutton-Brock, 99–118. Chicago: University of Chicago Press.

Howard, R. D., and A. G. Kluge. 1985. "Proximate mechanisms of sexual selection in wood frogs." *Evolution* 39:260–277.

Howard, R. D., and J. R. Young. 1998. "Individual variation in male vocal traits and female preferences in *Bufo americanus*." *Animal Behaviour* 55:1165–1179.

Humphries, W. J., and M. A. Sisson. 2012. "Long distance migrations, landscape use, and vulnerability to prescribed fire of the gopher frog (*Lithobates capito*)." *Journal of Herpetology* 46:665–670.

Jansen, K. P., A. P. Summers, and P. R. Delis. 2001. "Spadefoot toads (*Scaphiopus holbrookii holbrookii*) in an urban landscape: Effects of nonnatural substrates on burrowing in adults and juveniles." *Journal of Herpetology* 35:141–145.

Jensen, J. B. (ed.). 2008. *Amphibians and Reptiles of Georgia*. Athens: University of Georgia Press.

Jensen, J. B., M. A. Bailey, E. L. Blankenship, and C. D. Camp. 2003. "The relationship between breeding by the gopher frog, *Rana capito* (Ranidae: Anura), and rainfall." *American Midland Naturalist* 150:185–190.

Jones, K. S., and T. A. Tupper. 2015. "Fowler's toad (*Anaxyrus fowleri*) occupancy in the southern mid-Atlantic, USA." *Amphibian and Reptile Conservation* 9:24–33.

Kats, L. B., J. W. Petranka, and A. Sih. 1988. "Antipredator defenses and the persistence of amphibian larvae and fishes." *Ecology* 69:1865–1870.

Kime, N. M., S. S. Burmeister, and M. J. Ryan. 2004. "Female preferences for socially variable call characters in the cricket frog, *Acris crepitans*." *Animal Behaviour* 68:1391–1399.

Kinney, V. C., J. L. Heemeyer, A. P. Pessier, and M. J. Lannoo. 2011. "Seasonal pattern of *Batrachochytrium dendrobatidis* infection and mortality in *Lithobates areolatus*: Affirmation of Vredenburg's '10,000 zoospore rule.'" *PLOS One* 6:e16708.

Kirlin, M. S., M. M. Gooch, S. J. Price, and M. E. Dorcas. 2006. "Predictors of winter anuran calling in the North Carolina Piedmont." *Journal of the North Carolina Academy of Science* 122:10–18.

Kupferberg, S. J. 1997. "Bullfrog (*Rana catesbeiana*) invasion of a California river: The role of larval competition." *Ecology* 78:1736–1751.

Labanick, G. M. 1976. "Prey availability, consumption and selection in the Cricket Frog, *Acris crepitans* (Amphibia, Anura, Hylidae)." *Journal of Herpetology* 10:293–298.

LaFiandra, E. M., and K. J. Babbitt. 2004. "Predator induced phenotypic plasticity in the pine woods tree frog, *Hyla femoralis*: Necessary cues and the cost of development." *Oecologia* 138:350–359.

Lamb, T. 1984. "Amplexus displacement in the southern toad, *Bufo terrestris*." *Copeia* 1984:1023–1025.

Lamb, T., and J. C. Avise. 1986. "Directional introgression of mitochondrial DNA in a hybrid population of tree frogs: The influence of mating behavior." *Proceedings of the National Academy of Science* 83:2526–2530.

Lance, S. L., M. R. Erickson, R. W. Flynn, G. L. Mills, T. D. Tuberville, and D. E. Scott. 2012. "Effects of chronic copper exposure on development and survival in the southern leopard frog (*Lithobates [Rana] sphenocephalus*)." *Environmental Toxicology and Chemistry* 31:1587–1594.

Lance, S. L., R. W. Flynn, M. R. Erickson, and D. E. Scott. 2013. "Within- and among-population level differences in response to chronic copper exposure in southern toads, *Anaxyrus terrestris*." *Environmental Pollution* 177:135–142.

Lanctôt, C., C. Robertson, L. Navarro-Martín, C. Edge, S. D. Melvin, J. Houlahan, and V. L. Trudeau. 2013. "Effects of the glyphosate-based herbicide Roundup WeatherMax® on metamorphosis of wood frogs (*Lithobates sylvaticus*) in natural wetlands." *Aquatic Toxicology* 140–141:48–57.

Leary, C. J. 2001. "Evidence of convergent character displacement in release vocalizations of *Bufo fowleri* and *Bufo terrestris* (Anura: Bufonidae)." *Animal Behaviour* 61:431–438.

Leary, C. J., D. J. Fox, D. B. Shepard, and A. M. Garcia. 2005. "Body size, age, growth and alternative mating tactics in toads: Satellite males are smaller but not younger than calling males." *Animal Behaviour* 70:663–671.

Leary, C. J., and S. Harris. 2012. "Steroid hormone levels in calling males and males practicing alternative non-calling mating tactics in the green treefrog, *Hyla cinerea*." *Hormones and Behavior* 63:20–24.

Lehtinen, R. M., and S. M. Galatowitsch. 2001. "Colonization of restored wetlands by amphibians in Minnesota." *American Midland Naturalist* 145:388–396.

Lemmon, E. M. 2009. "Diversification of conspecific signals in sympatry: Geographic overlap drives multidimensional reproductive character displacement in frogs." *Evolution* 63:1155–1170.

Lemmon, E. M., A. R. Lemmon, J. T. Collins, J. A. Lee-Yaw, and D. C. Cannatella. 2007. "Phylogeny-based delimitation of species boundaries and contact zones in the trilling chorus frogs (*Pseudacris*)." *Molecular Phylogenetics and Evolution* 44:1068–1082.

Liner, A. E., L. L. Smith, S. W. Golladay, S. B. Castleberry, and J. W. Gibbons. 2008. "Amphibian distributions within three types of isolated wetlands in southwest Georgia." *American Midland Naturalist* 160:69–81.

Löding, H. P. 1922. "A preliminary catalogue of Alabama amphibians and reptiles." Museum Paper No. 5, Alabama Museum of Natural History. Geological Survey of Alabama, Tuscaloosa, AL.

Makowsky, R., J. Chesser, and L. J. Rissler. 2009. "A striking lack of genetic diversity across the wide-ranging amphibian *Gastrophryne carolinensis* (Anura: Microhylidae)." *Genetica* 135:169–183.

Martínez-Rivera, C. C., and H. C. Gerhardt. 2008. "Advertisement-call modification, male competition and female preference in the bird-voiced treefrog *Hyla avivoca*." *Behavioral Ecology and Sociobiology* 63:195–208.

Masta, S. E., B. K. Sullivan, T. Lamb, and E. J. Routman. 2002. "Molecular systematics, hybridization, and phylogeography of the *Bufo americanus* complex in eastern North America." *Molecular Phylogenetics and Evolution* 24:302–314.

McCallum, M. L., and S. E. Trauth. 2003. "A forty-three year museum study of northern cricket frog (*Acris crepitans*) abnormalities in Arkansas: Upward trends and distributions." *Journal of Wildlife Diseases* 39:522–528.

McCollum, S. A., and J. D. Leimberger. 1997. "Predator-induced morphological changes in an amphibian: Predation by dragonflies affects tadpole shape and color." *Oecologia* 109:615–621.

McConnell, R., T. McConnell, C. Guyer, and D. Laurencio. 2015. "*Eleutherodactylus cystignathoides* (Rio Grande Chirping Frog)." *Herpetological Review* 46:558–559.

McCrudden, C. M., M. Zhou, T. B. Chen, M. O'Rourke, B. Walker, D. G. Hirst, and C. Shaw. 2007. "The complex array of bradykinin-related peptides (BRPs) in the peptidome of pickerel frog (*Lithobates palustris*) skin secretion is the product of transcriptional economy." *Peptides* 28:1275–1281.

McKnight, T. L., and D. Hess. 2000. *Physical Geography: A Landscape Appreciation.* Upper Saddle River, NJ: Prentiss-Hall.

Means, D. B., and P. E. Moler. 1979. "The Pine Barrens Treefrog: Fire, seepage bogs, and management implications." In *Proceedings of the Rare and Endangered Wildlife Symposium*, ed. R. R. Odum and L. Landers, 77–83. Atlanta: Georgia Department of Natural Resources.

Mecham, J. S. 1960. "Introgressive hybridization between two southeastern treefrogs." *Evolution* 14:445–457.

Mendelson, J. R., III, K. T. Chase, and J. B. Murphy. 2015. "A review of the biology and literature of the Gulf Coast toad (*Incilius nebulifer*), native to Mexico and the United States." *Zootaxa* 3974:517–537.

Mendelson, J. R., III, D. C. Mulcahy, T. S. Williams, and J. W. Sites Jr. 2011. "A phylogeny and evolutionary natural history of Mesoamerican toads (Anura: Bufonidae: *Incilius*) based on morphology, life history, and molecular data." *Zootaxa* 3138:1–34.

Meshaka, W. E. 2011. "A runaway train in the making: The exotic amphibians, reptiles, turtles, and crocodilians of Florida." *Herpetological Conservation and Biology* 6:1–101.

Meshaka, W. E., B. P. Butterfield, and J. B. Hauge. 2004. *Exotic Amphibians and Reptiles of Florida.* Malabar, FL: Krieger.

Meshaka, W. E., and J. N. Layne. 2005. "Habitat relationships and seasonal activity of the greenhouse frog (*Eleutherodactylus planirostris*) in southern Florida." *Florida Scientist* 68:35–43.

Metts, B. S., K. A. Buhlmann, T. D. Tuberville, D. E. Scott, and W. A.

Hopkins. 2013. "Maternal transfer of contaminants and reduced reproductive success of southern toads (*Bufo [Anaxyrus] terrestris*) exposed to coal combustion waste." *Environmental Science and Technology* 47:2846–2853.

Micancin, J. P., and J. T. Mette. 2009. "Acoustic and morphological identification of the sympatric cricket frogs *Acris crepitans* and *A. gryllus* and the disappearance of *A. gryllus* near the edge of its range." *Zootaxa* 2076:1–36.

Miller, D., M. Gray, and A. Storfer. 2011. "Ecopathology of ranaviruses infecting amphibians." *Viruses* 3:2351–2373.

Moler, P. E. 1981. "Notes on *Hyla andersonii* in Florida and Alabama." *Journal of Herpetology* 15:441–444.

Moriarty, E. C., and D. C. Cannatella. 2004. "Phylogenetic relationships of the North American chorus frogs (*Pseudacris*: Hylidae)." *Molecular Phylogenetics and Evolution* 30:409–420.

Mount, R. H. 1975. *Reptiles and Amphibians of Alabama.* Auburn: Alabama Agricultural Experiment Station, Auburn University.

Murphy, C. G. 1992. "Nightly timing of chorusing by male barking treefrogs (*Hyla gratiosa*): The influence of female arrival and energy." *Copeia* 1992:333–347.

Murphy, C. G. 1994. "Chorus tenure of male barking tree frogs, *Hyla gratiosa.*" *Animal Behaviour* 48:763–777.

Murphy, C. G., and C. H. Gerhardt. 1996. "Evaluating the design of mate-choice experiments: The effects of amplexus on mate choice by female barking treefrogs, *Hyla gratiosa.*" *Animal Behaviour* 51:881–890.

Navarro-Martín, L., C. Lanctôt, P. Jackman, B. J. Park, and K. Doe. 2014. "Effects of glyphosate-based herbicides on survival, development, growth and sex ratios of wood frog (*Lithobates sylvaticus*) tadpoles. I: Chronic laboratory exposure to VisionMax®." *Aquatic Toxicology* 154:278–290.

Newman, C. E., and L. J. Rissler. 2011. "Phylogeographic analyses of the southern leopard frog: The impact of geography and climate on the distribution of genetic lineages vs. subspecies." *Molecular Ecology* 20:5295–5312.

Oldham, R. S., and H. C. Gerhardt. 1975. "Behavioral isolating mechanisms of the treefrogs *Hyla cinerea* and *H. gratiosa.*" *Copeia* 1975:223–231.

Ortiz-Santaliestra, E. M., and D. W. Sparting. 2007. "Alteration of larval development and metamorphosis by nitrate and perchlorate in southern leopard frogs (*Rana sphenocephala*)." *Archives of Environmental Contamination and Toxicology* 53:639–646.

Ospina, O. E., L. Tieu, J. J. Apodaca, and E. M. Lemmon. 2020. "Hidden diversity in the Mountain Chorus Frog (*Pseudacris brachyphona*) and the diagnosis of a new species of chorus frog in the southeastern United States." *Copeia* 108:778–795.

Palis, J. G. 1998. "Breeding biology of the gopher frog, *Rana capito*, in west Florida." *Journal of Herpetology* 32:217–223.

Parris, M. J., and D. R. Baud. 2004. "Interactive effects of a heavy metal and chytridiomycosis on gray treefrog larvae (*Hyla chrysoscelis*)." *Copeia* 2004:344–350.

Pauly, G. B., D. M. Hillis, and D. C. Cannatella. 2004. "The history of a Nearctic colonization: Molecular phylogenetics and biogeography of the Nearctic toads (*Bufo*)." *Evolution* 58:2517–2535.

Pauly, G. B., D. M. Hillis, and D. C. Cannatella. 2009. "Taxonomic freedom and the role of official lists of species names." *Herpetologica* 65:115–128.

Pearson, P. G. 1955. "Population ecology of the spadefoot toad, *Scaphiopus h. holbrooki* (Harlan)." *Ecological Monographs* 25:233–267.

Pechmann, J. H. K., D. E. Scott, R. D. Semlitsch, J. P. Caldwell, L. J. Vitt, and J. W. Gibbons. 1991. "Declining amphibian populations: The problem of separating human impacts from natural fluctuation." *Science* 253:892–895.

Perrill, S. A., H. C. Gerhardt, and R. Daniel. 1978. "Sexual parasitism in the Green Tree Frog (*Hyla cinerea*)." *Science* 200:1179–1180.

Perrill, S. A., and W. J. Shepherd. 1989. "Spatial distribution and male-male communication in the Northern Cricket Frog, *Acris crepitans blanchardi*." *Journal of Herpetology* 2:237–243.

Petranka, J. W., E. M. Harp, C. T. Holbrook, and J. A. Hamel. 2007. "Long-term persistence of amphibian populations in a restored wetland complex." *Biological Conservation* 138:371–380.

Pham, L., S. Boudreaux, S. Karhbet, B. Price, A. S. Ackleh, J. Carter, and N. Pal. 2007. "Population estimates of *Hyla cinerea* (Schneider) (Green Treefrog) in an urban environment." *Southeastern Naturalist* 6:203–216.

Phillips, B. L., G. P. Brown, and R. Shine. 2003. "Assessing the potential impact of cane toads on Australian snakes." *Conservation Biology* 17:1738–1747.

Phillips, C. A., R. A. Brandon, and E. O. Moll. 1999. *Field Guide to the Amphibians and Reptiles of Illinois*. Champaign: Illinois Natural History Survey.

Poole, K. G., and C. G. Murphy. 2007. "Preference of female barking treefrogs, *Hyla gratiosa*, for larger males: Univariate and composite tests." *Animal Behaviour* 73:513–524.

Pramuk, J. B., T. Robertson, J. W. Sites Jr., and B. P. Noonan. 2007. "Around the world in 10 million years: Biogeography of the nearly cosmopolitan true toads (Anura: Bufonidae)." *Global Ecology and Biogeography* 17:72–83.

Pullen, K. D., A. M. Best, and J. L. Ware. 2010. "Amphibian pathogen *Batrachochytrium dendrobatidis* prevalence is correlated with season and not urbanization in central Virginia." *Diseases of Aquatic Organisms* 91:9–16.

Punzo, F. 1992. "Dietary overlap and activity patterns in sympatric populations *Scaphiopus holbrooki* (Pelobatidae) and *Bufo terrestris* (Bufonidae)." *Florida Scientist* 55:38–44.

Pyron, R. A., and J. J. Weins. 2011. "A large-scale phylogeny of Amphibia including over 2800 species, and a revised classification of extant frogs, salamanders, and caecilians." *Molecular Phylogenetics and Evolution* 61:543–583.

Rage, J., and Z. Rocek. 1989. "Redescription of *Triadobatrachus massinoti* (Piveteau, 1936) an anuran amphibian from the early Triassic." *Palaeontographica* Abt. A 206:1–16.

Reeder, A. L., M. O. Ruiz, A. Pessier, L. E. Brown, J. M. Levengood, C. A. Phillips, M. B. Wheeler, R. E. Warner, and V. R. Beasley. 2005. "Intersexuality and the Cricket Frog decline: Historic and geographic trends." *Environmental Health Perspectives* 113:261–265.

Reeves, M. K., C. L. Dolph, H. Zimmer, R. S. Tjeerdema, and K. A. Trust. 2008. "Road proximity increases risk of skeletal abnormalities in wood frogs from National Wildlife Refuges in Alaska." *Environmental Health Perspectives* 116:1009–1014.

Regosin, J. V., B. S. Windmiller, and J. M. Reed. 2003. "Terrestrial habitat use and winter densities of the wood frog (*Rana sylvatica*)." *Journal of Herpetology* 37:390–394.

Resetarits, W. J., Jr. 1986. "Ecology of cave use by the frog, *Rana palustris*." *American Midland Naturalist* 116:256–266.

Richter, S. C., and R. E. Broughton. 2005. "Development and characteristics of polymorphic microsatellite DNA loci for the endangered dusky gopher frog, *Rana sevosa*, and two closely related species." *Molecular Ecology* 5:436–438.

Richter, S. C., B. I. Crother, and R. E. Broughton. 2009. "Genetic consequences of population reduction and geographic isolation in the critically endangered frog, *Rana sevosa*." *Copeia* 2009:799–806.

Richter, S. C., E. M. O'Neill, S. O. Nunziata, A. Rumments, E. S. Gustin, J. E. Young, and B. I. Crother. 2014. "Cryptic diversity and conservation of gopher frogs across southeastern United States." *Copeia* 2014:231–237.

Richter, S. C., and R. A. Seigel. 2002. "Annual variation in the population ecology of the endangered gopher frog, *Rana sevosa* Goin and Netting." *Copeia* 2002:962–972.

Richter, S. C., J. E. Young, G. N. Johnson, and R. A. Seigel. 2003. "Stochastic variation in reproductive success of a rare frog, *Rana sevosa*: Implications for conservation and for monitoring amphibian populations." *Biological Conservation* 111:171–177.

Richter, S. C., J. E. Young, R. A. Seigel, and G. N. Johnson. 2001. "Postbreeding movements of the dark gopher frog, *Rana sevosa* Goin and Netting: Implications for conservation and management." *Journal of Herpetology* 35:316–321.

Rieger, J. F., C. A. Binckley, and W. J. Resetarits Jr. 2004. "Larval

performance and oviposition site preference along a predation gradient."
Ecology 85:2094–2099.

Rittenhouse, T. A. G., E. B. Harper, L. R. Rehard, and R. D. Semlitsch. 2008.
"The role of microhabitats in the desiccation and survival of anurans in
recently harvested oak-hickory forest." *Copeia* 2008:807–814.

Rizkalla, C. E. 2010. "Increasing detection of *Batrachochytrium dendrobatidis*
in central Florida, USA." *Herpetological Review* 41:180–181.

Rollins, A. W., J. E. Copeland, H. Barker, and D. Satterfield. 2013. "The distribution of *Batrachochytrium dendrobatidis* across the southern Appalachian
states, USA." *Mycosphere* 4(2):250–254.

Romagosa, C. A., C. Guyer, and M. C. Wooten. 2009. "Contribution of the
live-vertebrate trade toward taxonomic homogenization." *Conservation
Biology* 23:1001–1007.

Rowe, C. L., W. A. Hopkins, and V. R. Coffman. 2001. "Failed recruitment
of Southern Toads (*Bufo terrestris*) in a trace element-contaminated
breeding habitat: Direct and indirect effects that may lead to a local population sink." *Archives of Environmental Contamination and Toxicology*
40:399–405.

Roznik, E. A., and S. A. Johnson. 2009. "Burrow use and survival of newly
metamorphosed gopher frogs (*Rana capito*)." *Journal of Herpetology*
43:431–437.

Roznik, E. A., S. A. Johnson, C. H. Greenberg, and G. W. Tanner. 2009. "Terrestrial movements and habitat use of gopher frogs in longleaf pine forest:
A comparative study of juveniles and adults." *Forest Ecology and Management* 259:187–194.

Ruthig, G. R. 2009. "Water molds of the genera *Saprolegnia* and *Leptolegnia*
are pathogenic to the North American frogs *Rana catesbeiana* and *Pseudacris crucifer*, respectively." *Diseases of Aquatic Organisms* 84:173–178.

Ryan, M. J. 1980. "The reproductive behavior of the bullfrog (*Rana catesbeiana*)." *Copeia* 1980:108–114.

Ryan, M. J., K. M. Warkentin, B. E. McClelland, and W. Wilczynski. 1995.
"Fluctuating asymmetries and advertisement call variation in the cricket
frog, *Acris crepitans*." *Behavioral Ecology* 6:124–131.

Saenz, D., B. T. Kavanagh, and M. Kwiatkowski. 2010. "*Batrachochytrium dendrobatidis* detected in amphibians from national forests in eastern Texas,
USA." *Herpetological Review* 41:47–49.

Schiesari, L., E. E. Werner, and G. W. Kling. 2009. "Carnivory and resource-
based niche differentiation in anuran larvae: Implication for food web and
experimental ecology." *Freshwater Biology* 54:572–586.

Schotz, A., and M. Barbour. 2009. "Ecological assessment and terrestrial vertebrate surveys for Black Belt Prairies in Alabama." www.outdooralabama
.com.

Schwartz, A., and R. W. Henderson. 1991. *Amphibians and Reptiles of the West Indies: Descriptions, Distributions, and Natural History.* Gainesville: University of Florida Press.

Seigel, R. A., and C. K. Dodd. 2002. "Translocation of amphibians: Proven management method or experimental technique?" *Conservation Biology* 16:552–554.

Semlitsch, R. D., and J. P. Caldwell. 1982. "Effects of density on growth, metamorphosis, and survivorship in tadpoles of *Scaphiopus holbrooki.*" *Ecology* 63:905–911.

Shelton-Nix, E. (ed.). 2017. *Alabama Wildlife*, vol. 5. Tuscaloosa: University of Alabama Press.

Sin, Y., M. Zhou, W. Chen, L. Wang, T. Chen, B. Walker, and C. Shaw. 2008. "Skin bradykinin-related peptides (BRPs) and their biosynthetic precursors (kinogens): Comparisons between various taxa of Chinese and North American ranid frogs." *Peptides* 29:393–403.

Smith, L. L., and C. K. Dodd. 2003. "Wildlife mortality on U.S. Highway 44 across Paynes Prairie, Alachua County, Florida." *Florida Scientist* 66:128–140.

Smith, P. W. 1961. *The Amphibians and Reptiles of Illinois.* Illinois Natural History Survey Bulletin 28. Urbana: Natural History Survey Division, Department of Registration and Education, State of Illinois.

Soltis, D. E., A. B. Morris, J. S. McLachlan, P. S. Manos, and P. S. Soltis. 2006. "Comparative phylogeography of unglaciated eastern North America." *Molecular Ecology* 15:4261–4293.

Steen, D. A., C. J. W. McClure, and S. P. Graham. 2013. "Relative influence of weather and season on anuran calling." *Canadian Journal of Zoology* 91:462–467.

Steen, D. A., A. E. Rall McGee, S. M. Hermann, J. A. Stiles, S. H. Stiles, and C. Guyer. 2010. "Effects of forest management on amphibians and reptiles: Generalist species obscure trends among native forest associates." *Open Environmental Sciences* 4:24–30.

Stiles, R. M., M. J. Sieggreen, R. A. Johnson, K. Pratt, M. Vassallo, M. Andrus, M. Perry, J. W. Swan, and M. J. Lannoo. 2016. "Captive-rearing state endangered crawfish frogs *Lithobates areolatus* from Indiana, USA." *Conservation Evidence* 13:7–11.

Strojny, C. A., and M. L. Hunter Jr. 2010. "Relative abundance of amphibians in forest canopy gaps of natural origin *vs.* timber harvest origin." *Animal Biodiversity and Conservation* 33(1):1–13.

Sullivan, B. K. 1992. "Sexual selection and calling behavior in the American toad (*Bufo americanus*)." *Copeia* 1992:1–7.

Sutton, W. B., M. J. Gray, R. H. Hardman, R. P. Wilkes, A. J. Kouba, and D. L.

Miller. 2014. "High susceptibility of the endangered Dusky Gopher Frog to ranavirus." *Disease of Aquatic Organisms* 112:9–16.

Terrell, V. C. K., N. J. Engbrecht, A. P. Pessier, and M. J. Lannoo. 2014. "Drought reduces chytrid fungus (*Batrachochytrium dendrobatidis*) infection intensity and mortality but not prevalence in adult Crawfish Frogs (*Lithobates areolatus*)." *Journal of Wildlife Diseases* 50:56–62.

Timm, B. C., K. McGarigal, and R. P. Cook. 2014. "Upland movement patterns and habitat selection of adult Eastern Spadefoots (*Scaphiopus holbrookii*) at Cape Cod National Seashore." *Journal of Herpetology* 48:84–97.

Todd, B. D., T. M. Luhring, B. B. Rothermel, and J. W. Gibbons. 2009. "Effects of forest removal on amphibian migrations: Implications for habitat and landscape connectivity." *Journal of Applied Ecology* 46:554–561.

Todd, B. D., and B. B. Rothermel. 2006. "Assessing quality of clearcut habitats for amphibians: Effects on abundance versus vital rates in the Southern Toad (*Bufo terrestris*)." *Biological Conservation* 133:178–185.

Todd, B. D., and C. T. Winne. 2006. "Ontogenetic and interspecific variation in timing of movement and responses to climatic factors during migrations by pond-breeding amphibians." *Canadian Journal of Zoology* 84:715–722.

Todd, M. J., R. R. Cocklin, and M. E. Dorcas. 2003. "Temporal and spatial variation in anuran calling activity in the western piedmont of North Carolina." *Journal of the North Carolina Academy of Science* 119:103–110.

Travis, J., and J. C. Trexler. 1984. "Investigation on the control of the color polymorphism in *Pseudacris ornata*." *Herpetologica* 40:252–257.

Traylor, R. C., B. W. Buchanon, and J. L. Doherty. 2007. "Sexual selection in the Squirrel Treefrog *Hyla squirella*: The role of multimodal cue assessment in female choice." *Animal Behaviour* 74:1753–1763.

US Fish and Wildlife Service. 2015. Dusky Gopher Frog (*Rana sevosa*) Recovery Plan. Atlanta, GA. https://ecos.fws.gov/ecp/.

US Fish and Wildlife Service. 2019. Dusky Gopher Frog (*Rana sevosa*) Recovery Plan Amendment. Atlanta, GA. https://ecos.fws.gov/ecp/.

Vasconcelos, D., and A. J. K. Calhoun. 2004. "Movement patterns of adult and juvenile *Rana sylvatica* (LeConte) and *Ambystoma maculatum* (Shaw) in three restored seasonal pools in Maine." *Journal of Herpetology* 38:5521–561.

Vogel, L. S., and J. H. K. Pechmann. 2010. "Response of Fowler's Toad (*Anaxyrus fowleri*) to competition and hydroperiod in the presence of the invasive Coastal Plain Toad (*Incilius nebulifer*)." *Journal of Herpetology* 44:382–389.

Volpe, E. P., and J. L. Dobie. 1959. "The larva of the Oak Toad, *Bufo quercicus* Holbrook." *Tulane Studies in Zoology* 7:145–152.

Waldman, B. 1982. "Adaptive significance of communal oviposition in Wood Frogs (*Rana sylvatica*)." *Behavioral Ecology and Sociobiology* 10:169–174.

Walpole, A. A., J. Bowman, D. C. Tozer, and D. S. Badzinski. 2012. "Community-level response to climate change: Shifts in anuran calling phenology." *Herpetological Conservation and Biology* 7:249–257.

Warwick, A. R., J. Travis, and E. M. Lemmon. 2015. "Geographic variation in the Pine Barrens Treefrog (*Hyla andersonii*): Concordance of genetic, morphometric, and acoustic signal data." *Molecular Ecology* 24:3281–3298.

Weir, L. A., J. A. Royale, K. D. Gazenski, and O. Villena. 2014. "Northeast regional and state trends in anuran occupancy from calling survey data (2001–2011) from the North American Amphibian Monitoring Program." *Herpetological Conservation and Biology* 9:223–245.

Weir, S. W., S. Yu, C. J. Salice. 2012. "Acute toxicity of herbicide formulations and chronic toxicity of technical-grade Trifluralin to larval Green Frogs (*Lithobates clamitans*)." *Environmental Toxicology and Chemistry* 31:2029–2034.

Welch, A. M., R. D. Semlitsch, and H. C. Gerhardt. 1998. "Call duration as an indicator of genetic quality in male gray tree frogs." *Science* 280:1928–1930.

Wells, K. D. 1977. "Territoriality and mating success in the Green Frog (*Rana clamitans*)." *Ecology* 58:1053–1066.

Wells, K. D. 1978. "Territoriality in the Green Frog (*Rana clamitans*): Vocalizations and agonistic behavior." *Animal Behaviour* 26:1051–1063.

Werner, E. E., G. A. Wellborn, and M. A. McPeek. 1995. "Diet composition in post-metamorphic bullfrogs and green frogs: Implications for interspecific predation and competition." *Journal of Herpetology* 29:600–607.

Wiens, J. J. 2007. "Global patterns of diversification and species richness in amphibians." *American Naturalist* 170:S86–S106.

Wiens, J. J., J. W. Fetzner, C. L. Parkinson, and T. W. Reeder. 2006. "Hylid frog phylogeny and sampling strategies for speciose clades." *Systematic Biology* 54(5):719–748.

Wilbur, H. M. 1977. "Density dependent aspects of growth and metamorphosis in *Bufo americanus*." *Ecology* 58:196–200.

Wilbur, H. M., and J. E. Fauth. 1990. "Experimental aquatic food webs: Interactions between two predators and two prey." *American Naturalist* 135:176–204.

Wilbur, H. M., D. I. Rubenstein, and L. Fairchild. 1978. "Sexual selection in toads: The roles of female choice and male body size." *Evolution* 32:264–270.

Williams, P. J., J. R. Robb, and D. R. Karns. 2012. "Habitat selection by Crawfish Frogs (*Lithobates areolatus*) in a large mixed grassland/forest habitat." *Journal of Herpetology* 46:682–688.

Wood, L., and A. M. Welch. 2015. "Assessment of interactive effects of elevated salinity and three pesticides on life history and behavior of Southern Toads (*Anaxyrus terrestris*) tadpoles." *Environmental Toxicology and Chemistry* 34:667–676.

Woods, K. V., J. D. Nichols, H. F. Percival, and J. E. Hines. 1998. "Size-sex variation in survival rates and abundance of Pig Frogs, *Rana grylio*, in northern Florida wetlands." *Journal of Herpetology* 32:527–535.

Wright, A. H., and A. A. Wright. 1949. *Handbook of Frogs and Toads of the United States and Canada.* Third Edition, Comstock Publisher, Ithaca, NY.

Yuan, Z., W. Zhou, X. Chen, N. A. Poyarkov Jr., H. Chen, N. Jang-Liaw, W. Chou, N. J. Matzke, K. Iizuka, M. Min, S. L. Kuzmin, Y. Zhang, D. C. Cannatella, D. M. Hillis, and J. Che. 2016. "Spatiotemporal diversification of the True Frogs (Genus *Rana*): A historical framework for a widely studied group of model organisms." *Systematic Biology* 65:824–842.

About the Authors

CRAIG GUYER was born on August 6, 1952, in Los Angeles, California, to parents who migrated to several towns across southern California before settling on a ranch on the outskirts of Oceanside. Although now decimated by high-density housing, during his childhood this area was remote, covered with coastal sage scrub vegetation, and home to healthy populations of amphibians and reptiles. So, when he reached third grade, a stage when many children develop a fascination with wildlife, his backyard provided daily opportunities for exploration, an interest that he has never outgrown. Craig attended Humboldt State University in Northern California, graduating with a Zoology degree in 1975. He then moved to Idaho State University and in 1978 completed a master's degree that examined homing behavior in sagebrush lizards and short-horned lizards. He entered the lab of Dr. Jay M. Savage, then at the University of Southern California but who later chaired the Department of Biological Sciences at the University of Miami, for doctoral work. This career step introduced Craig to the diverse herpetofauna of Costa Rica, where he continues to maintain active research. He accepted a position at Auburn University in 1987, replacing the retired Robert H. Mount as curator of herpetology. Over the years, the Guyer lab has maintained the premiere collection of Alabama amphibians and reptiles, provided research vital to the conservation of Alabama's amphibians and reptiles, and taught undergraduate students about global patterns of vertebrate biodiversity. He is the 2013 recipient of the Meritorious Teaching Award given by the Society for the Study of Amphibians and Reptiles and was the 2013–2016 Scharnagel Professor of Biological Sciences in the College of Science and Mathematics at Auburn University. He retired in 2016.

MARK A. BAILEY was born on July 20, 1961, in Birmingham, Alabama. His fascination with reptiles and amphibians began around age nine when he was given Hobart M. Smith's *Snakes as Pets*, and his formative years were spent roaming the woods of the upper Black Warrior River watershed, bringing home more "pets" than a conservation-minded herpetologist should probably confess. He was fortunate to have parents who tolerated and supported his interests. He attended public and private schools in Huffman, Palmerdale, and Pinson, Alabama. After graduating from high school in 1979, he entered Auburn University where he completed a BS degree in Biology in 1984 and an MS degree in Zoology in 1988, with Robert H. Mount as his major professor.

From 1986 to 1988, he was assistant curator of the Auburn University Museum's herpetology collection. Immediately following graduation, he worked as a Biological Technician at Conecuh National Forest, and in 1989, he was hired as Zoologist for the Nature Conservancy's new Alabama Natural Heritage Program. There, he conducted statewide reptile and amphibian field surveys, organized conservation plans, and initiated the Alabama Herp Atlas Project. In 1998, he left the Heritage Program and with his wife, Karan, founded Conservation Southeast, a consulting firm they continue to operate from Andalusia. He has served on the advisory board of the Auburn University School of Forestry and Wildlife Science and is a past president of the Alabama Chapter of the Wildlife Society. He has served as the Alabama state representative to the Gopher Tortoise Council, and he continues to serve as a director of the Alabama Wildlife Federation. He received the Governor's Conservation Achievement Award for Wildlife Biologist of the Year in 2007. Mark has been involved in Partners for Amphibian and Reptile Conservation (PARC) and the Alabama chapter (ALAPARC) since their inception and was lead author on the 2006 PARC publication, *Management Guidelines for Amphibians and Reptiles of the Southeastern United States.*

Index

Page numbers in italics refer to figures.

Florida, 4, 5, 6, 11, 17, 67, 78, 105, 115, 127, 130, 138, 141, 144, 157
Florida Leopard Frog. See *Lithobates sphenocephalus sphenocephalus*
Florida Panhandle, 116, 157, 163, 200, 201
Fort Morgan peninsula, 17
Fort Rucker, 63
Fowler's Toad. See *Anaxyrus fowleri*
Franklin County, 24

Gastrophryne: genus, x, 13, 80; *carolinensis*, x, 3, 32, 41, 42, 80, 80–83, 197
Geneva County, 6, 198
Geneva State Forest, 63
Georgia, 14, 20, 23, 28, 94, 95, 104, 115, 127, 163
Glen's Pond, 119
glyphosate, 97, 104
Gopher Frog. See *Lithobates capito*
Gopher Tortoise. See *Gopherus polyphemus*
Gopherus polyphemus, 5, 113, 141, 213
Great Caribbean Landfrogs. *See* Eleutherodactylidae
Green Frog. See *Lithobates clamitans*
Greenhouse Frog. See *Eleutherodactylus planirostris*
Green Treefrog. See *Dryophytes cinerea*
Gulf Coast Toad. See *Incilius nebulifer*
Gulf of Mexico, 1, 7, 10–11
Gulf State Park, 130

Hale County, 122
Hemisotidae, 79
Highland Rim, ix, 25, 32, 32, 206
Holarctic Treefrogs. See *Dryophytes*
Honduras, 80
Horseshoe Bend National Military Park, 96
Houston County, 92, 129, 161, 163
Hylidae, x, xi, 1, 40, 44, 145–47
Hyperoliidae, 79
Hypopachus, 80

Imperata cylindrica, 49
Incilius: genus, x, 51, 53, 57; *nebulifer*, x, 4, 53, 53–56, 67
indigenous species, ix, 2–4
introduced species, ix, 4–6

Jackson County, 28, 29, 30, 74, 98, 150, 155, 175, 178
Juniperus virginiana, 3

Lake Jackson, 17
Lamar County, 77
Lauderdale County, 32, 70, 133, 211
Lawrence County, 29, 32, 77
Lee County, 5, 23, 25, 77, 136
leptodactyliform, 145
Leptolegnia, 136, 173
Liquidambar styraciflua, 169
Limestone County, 32, 205
Lithobates: genus, xi, 43, 86–136; *areolatus circulosus*, xi, 3, 23, 90, *121*, 121–24; *capito*, xi, 4, 12, 37, 89, 90, *111*, 111–15; *catesbeianus*, xi, 4, 6, 40, 87, 91, 92, *132*, 133, 132–36; *clamitans*, xi, 4, 88, *101*, 102, 101–4; *grylio*, xi, 4, 19, 91, *128*, 128–31; *heckscheri*, xi, 4, 90, *125*, 125–27; *palustris*, xi, 4, 21, 88, *98*, 98–100; *sevosus*, xi, 4, *116*, 116–20; *sphenocephalus*, xi, 4, 74, 89, 105; *sphenocephalus sphenocephalus*, xi, 4, 89, *106*, 106–10; *sphenocephalus utricularius*, xi, 4, 89, 107, 106–10; *sylvaticus*, xi, 4, 27, 87, 93, 93–97
Litoria caerulea, 5
Little Grass Frog. See *Pseudacris ocularis*
Little Mountain, 29
Loblolly Pine. See *Pinus taeda*
Longleaf Pine. See *Pinus palustris*
Lookout Mountain, 28, 74
Louisiana, 5, 10, 49, 54, 56, 144, 157
Lower Coastal Plain, 20

Macon County, 19, 23
Madagascar, 51